DOMINANCE
AND
DELUSION

Why We Do The Things We Do

MARC A. CURTIS

Copyright © 2023 Marc A. Curtis
All rights reserved
First Edition

Fulton Books
Meadville, PA

Published by Fulton Books 2023

ISBN 979-8-88731-589-8 (paperback)
ISBN: 979-8-89221-166-6 (hardcover)
ISBN 979-8-88731-590-4 (digital)

Printed in the United States of America

INTRODUCTION

This is a book of whys. They are simple whys, relevant whys, and necessary whys. Why do humans behave as they do? Why do humans think as they do? Why do humans believe as they do? Some of these beliefs are so silly, so absurd, so ludicrous—in and of themselves—that they are unbelievable. Yet we believe, and the question must be why.

Why, for example, would anyone believe in astrology? We have the birth dates of many people, both famous and not so. There is absolutely no pattern to the dates or times of these births. Not the slightest sign of clustering to show even the tiniest indication that birth time was all or even a small part of one's life cycle.

Besides, a more logical point at which one would expect astrological intercession is conception. This is, after all, an exact time—a specific day, hour, minute, and second. But except for petri dishes, such knowledge is unavailable, so birth time, even though it can vary, is the basis for modern astrological signs. And so we believe in the unbelievable, content that we know the unknowable, and proud that we are born in the sign of the lion or the ram or the virgin. This is delusion on a grand scale.

There is yet another aspect of astrology that must be considered. How does it actually work? How is our birth impacted by the stars? Do we, at birth, give off some type of radiation that is noted and replied to by the stars? The only problem with this concept is that nothing can travel faster than the speed of light. It would take eighty years for a round trip to a star forty light-years away. By then, our future would be over.

Perhaps the stars emit a ray or a beam that registers upon the newborn and fixes its future. If such is the case, then we ought to be able to detect such a beam or ray. Further, we ought to be able to record its variations as different individuals are born. And we ought to be able to analyze, understand, and duplicate the beam or ray to make us all healthy, wealthy, and wise. Don't think it's gonna happen.

Our delusions are many and varied. Yet all stem from a single basic human misconception. We believe that we are special. We believe we are unique, one of a kind, different, extraordinary; and so it is necessary that we demonstrate this condition by validating its veracity. The stars move for us. The planets position themselves for our benefit. The universe applies its total concentration to the birth of each and every one of us. That's how special we are.

We believe in ESP, UFOs, and a host of other entities with different acronyms—each special, each kept from the general public by a secretive and unscrupulous bureaucracy. We believe that our fate is in the cards and that Miss Cleo can pinpoint the reasons for our ups and downs. And Miss Cleo was big business. The company she was part of was recently forced to give up $500 million in billings because of false and defective advertising. A lot of people gave Miss Cleo a call and got fleeced.

Why do so many of us believe that we are special? Different? Unique? Why is it so hard for most of us to accept the fact that we are evolved animals? Why do so many of us prefer dreams and delusions to reality? Should we rationalize the impossible in an attempt to confirm the contention that we are special, or is it better to discover the truth and know that we are not?

There is danger in rationalizing the impossible, and we fail utterly to see that.

Why do we have war? Why does capitalism work? Why doesn't socialism work? Why did North America (excluding Mexico) develop so differently from South and Central America (including Mexico)? Both continents were discovered at the same time and were settled by Europeans. Today, the United States is the strongest, safest, wealthiest country in the world while Central and South America, including Mexico, wallow in poverty and chaos. And the question must be why. Until very recently, all these countries were dictatorships while the United States was never a dictatorship. This has a great deal to do with why there is a difference.

Why do we have dictators? Why is Africa, today, a bloody mess of dictatorships, filled with killing and hate, greed and corruption? The answer is simple, and the European colonists had nothing to do with it. Why do we have dictatorships in Southeast Asia, in the Middle East, and in China? We recently passed a little-noted but rather important threshold. It seems that for the first time in the world today, there are more people who are free than there are in chains. But the question still must be why is anyone anywhere subjected to a dictatorship?

In the United States, one of the most common impossibilities that we try to rationalize is the unsupported assertion that the Christian Bible is, in fact, a reasonable explanation of how we got here and why. This patently false absurdity is called creationism. Because it is such a widespread and pervasive delusion, it will be dealt with in some detail in this volume.

While researching the premise for this manuscript, I have had an opportunity to read creationist literature and the scientific refutation of this nonsense. In all this research, not once have I come across anything that would give some indication as to why anyone would believe such a patently false absurdity, and very little on why it is so important that we have an accurate and rational explanation of human behavior. This is really the key. If we are to address the ills of mankind, we must proceed from reality rather than from dreams.

How powerful is creationism as a force for blinding mankind? Ask Kurt Wise. Mr. Wise is one very smart man with a degree in geophysics from the University of Chicago and a PhD in geology from Harvard. Mr. Wise is a creationist. In spite of his learning and, even more important, in spite of his observations, he believes that the biblical explanation is the only way to go. In fact, Mr. Wise is reported to have said, "If all the evidence in the universe turns against creationism, I would be the first to admit it, but I would still be a creationist."

Now this is delusion on a grand scale. The geologic column is the very thing that totally gives the lie to creationism. It is there, in place, telling stories for intelligent, rational, unbiased individuals. These stories scream evolution and unequivocally deny creation. But here is an individual able to wall off sections of his mind and believe even if all the evidence is to the contrary.

And the question still must be why. Why would an intelligent, educated individual clutch this delusion to his bosom? The answer is actually as mundane and as delusional as creationism.

This is the way to get into heaven. By believing, one is saved. Questioning leads to hell, and that is motivation enough for most.

Harmless, you say. What harm can come from believing? Well, in Mr. Wise's case, you are right. But you might be interested to know that the perpetrators of the 9/11 mass murder did what they did in the name of religion and for a reward of seventy-two virgins.

Yes, that's right. Each of those idiots who hijacked the planes and flew them into the World Trade Center did so because they had been promised the services of seventy-two virgins in heaven. Islamic fascination with virginity will be dealt with in more detail later. Suffice to say, such fascination runs deep in the Muslim world and profoundly influences the treatment of females throughout Islamic countries.

Getting back to the virgins. It does not seem to me that these guys have really thought the whole virginity concept through. Not meaning to be disrespectful to the ladies, but good sex is as much experience as it is ability. Virgins with little or no experience just lie there. After the tenth virgin, the whole concept begins to pall; after the fifteenth or twentieth, the job becomes a chore—on the same plane as mucking out stables; and, I suspect, virgins thirty through seventy-two remain in perpetuity just that, virgins.

They are a bit silly and rather quaint, some of these beliefs, and mostly harmless, except for the human element. It is, on the other hand, no coincidence that inhuman acts are committed only by humans. And that is the rub. I don't care what these guys believe; but when they say that they are right and everybody else is wrong, when they say they have not just a responsibility to convert the great unwashed masses but a mandate to do so, I get upset. This has been the story of God and mankind down through time. I am right and you are wrong, and I have the right to convert you or kill you. And the question still must be why.

Here's an interesting why. Why are some people homosexuals? Homosexuality is genetic, and homosexuals rarely have children. How then is the genetic predisposition for homosexuality passed on? There are reasons we behave as we do, and these reasons are rooted in our genetic past. We will not come to a better understanding of why we do the things we do by ducking into the closet of delusion. "It is that way because God did it that way" is not an explanation. Nor can we begin to address the problems of human behavior based on such an explanation.

Why do pundits maintain that we are monogamous when, in fact, 85 percent of the world's cultures are polygamous? Additionally, all the cultures in the world experience adultery, prostitution, and promiscuity. "Well," the pundits say, "we are partially monogamous." That is like saying a little bit pregnant or slightly dead. We either are or are not monogamous. The evidence, including our behavior, that we are not is overpowering and will be spelled out in detail. We will also delve into the reasons monogamy has been imposed upon the female by the male phylogenist. The reasons for this state of affairs have nothing to do with sex and have everything to do with dominance.

Why do we have war? Why are some individuals virtual dynamos of aggression while others are as pacifist as toast? Why is it that honest, just, and nice individuals can easily kill others in the heat and carnage of war where the enemy is clearly designated then go back to civilian life affected to a greater or lesser degree by the experience but not adversely influenced by the killing? Why is it so easy to get the population of a country to go to war? Germany had no problem putting together several armies during World War II. Neither did Japan. Nobody questioned these actions, certainly not enough to put any kind of brake on the mad dreams of Hitler or Tojo. Perhaps if these people had questioned their leader's behavior, they would not have suffered the retribution that they did. Leaders cannot lead followers who will not follow is a truism that needs to be understood by all the citizens of the world. It is also true that we get exactly the kind of government that we deserve.

Why do we have crime? Is it really the fact that some mother didn't breastfeed or that some male looked sideways at a vulnerable young child? Is it really true that if we are flooded with love during our formative years, we will be nice, kind, altruistic, honest, and loving? Interestingly enough for some, this is indeed true. Perhaps even for a majority of individuals, it is true. But for everyone? No. And the reasons are buried in our evolutionary past. You see, at one time, crime really did pay.

Why doesn't rehabilitation of criminals work? Actually, it does—sorta, kinda. Just enough to continue to give us hope that all will become susceptible, and not just the measly few we see today. Rehabilitation doesn't work very well because some individuals are wired to act "normal," and we find that offensive.

Why do so many teenage males form gangs? Why are so many of these gangs Hispanic and Black? What is so attractive about rebelling against what we euphemistically call societal norms? Is it possible that these individuals are acting normally? Is it possible that they are behaving in accordance with millions of years of genetically selected behavior patterns? If this is true, then society will not easily be able to rid the world of gangs.

This is the main reason such explanations are rejected by the leftist liberal elite. If teenage bonding patterns among males are genetically influenced, then all their sophisms about nice, kind, loving, gentle, caring, and compassionate upbringing go right out the window. In short, they lose their rationale for the perfectibility of mankind.

Why are there so many dictatorships in the world today? Why has the history of the world been a history of dictatorships? They have been called king, emperor, caesar; and one dictator referred to himself as El Maximo Leader. Why should such individuals exist in the first place? Why is it so important for social groups to establish lines of authority? Why is it that every culture ever studied consists of a dominance system? Why is it that every communist dictatorship that has ever been established has had to build walls around itself to keep people in? Why is it that capitalist democracies have to build walls around themselves to keep people out? There is a message here, and it has nothing to do with the welfare of mankind and everything to do with the dominance of men.

Speaking of communism, why doesn't it work? If we are really kind, caring, altruistic, naturally moral, and pacifist, then communism should work. Each individual should gladly give of himself to the extent that he is able and take only what is needed to survive. But it doesn't work that way. Every communist country in the world has become a dictatorship. Some still are. Not once has communism been able to demonstrate that it is superior to capitalism. And yet, if we are really nice, communism should work. So the question remains why.

On a lighter note, why do pubescent females go gaga over rock stars? Why do we find sporting contests so entertaining? Why do sports stars find it so easy to "score"? I'll bet Magic Johnson has asked himself that question a million times. Why do we gamble? Take drugs? Risk our lives with everything from bungee jumping to mountain climbing? In short, why do so many of us act the way we do? Why do we have risk-takers, CEOs, criminals, scientists, and, most of all, politicians?

There are answers to all these whys. They are simple, relatively straightforward, and, most importantly, evolutionary in origin. There are two fundamental characteristics that define the human animal. The first is dominance. All social animals form dominance systems for good, sound evolutionary reasons. We are social animals, and we form dominance systems. The other defining characteristic of the human animal is its ability to delude itself. This is a function of intelligence, and it is the single most serious flaw in the human animal. Dominance drives us—at least some of us—and delusion blinds most of us. That is what this book is all about.

CHAPTER 1

We are exceedingly fond of ourselves. We really, really like us. We think highly of the human race. We are the height of evolution in the eyes of many scientists, and we are made in the image of God, according to many religionists. We are special, unique, different, and the most important species on Earth. We are the culmination of evolution, the goal of Mother Nature, a perfect species.

This is delusion. Yes, we are different. Yes, we are unlike most animals. We are bipedal, hairless, and, some say, intelligent. But to imply that this difference entitles us to claim a total disconnect from the rest of biological life is a delusion. We are evolved animals. Much of our behavior is impacted by this fact. We will not truly begin to understand the human animal until we recognize this.

Our delusions are many and diverse. Some of us believe in ghosts. Some of us believe in psychics. Many of us believe in mental telepathy. Some even believe in haunted houses. We believe in ESP. We believe in life after death and "crossing over." We believe that we can come back from death, or have already come back from death, and will be able to do so again. All this without the slightest shred of proof. We are the victims of delusion.

We believe in good luck and bad luck and influencing luck. If one blows on the dice, that brings good luck. If one walks under a ladder, that is bad luck. Black cats, spilt salt, and broken mirrors are all bad luck. Four-leaf clovers are good luck. A rabbit's foot is thought to bring good luck but, as often pointed out, not for the rabbit, which at one time had four of these supposedly good luck charms. We believe because we want to, because we think we can influence our future, when, in fact, there is absolutely no evidence we can do so.

Delusion is not limited to life and luck. Some pretty intelligent and well-reasoned people also harbor delusion. This is the story of two skulls—one real, one a fake. The fake was immediately clutched to our anthropological bosom because it conformed to the then popular delusion that intelligence preceded looks. The real skull was rejected because it did not conform to our perception of human evolution. The real skull consisted of a brain case not much bigger than a chimp's but with a face obviously on the road to human shape. Since anthropologists of the time were under the brainy-ape delusion, which the fake skull supported, the real skull was rejected, and the fake skull accepted.

In 1912, some skull fragments were found in a gravel pit in Piltdown Sussex, England. The finds consisted of a partial skullcap and a partial lower jaw. The skullcap did not include any face bones, and the jaw was incomplete to the extent that it had neither chin nor jaw joint. The skullcap was essentially modern while the jaw was mostly ape with a few human features. It was a good but not perfect forgery. Clues were available and, in hindsight, obvious. But English gentlemen do not commit fraud, and so fraud was never entertained.

The real skull was rejected for the same reason the fake was accepted. It had a small brain while its facial features were decidedly more human and less like that of an ape. The animal also appeared to be bipedal. The fossil find was discovered in southern Africa, not England, and by a provincial anatomist, not an English gentleman. So the real was rejected while the fake was accepted for over forty years until modern test methods demonstrated that brain-first evolution was an English gentleman's delusion.

Such imbroglios, while uncommon, do occur. And in the long run, they have little effect on our understanding of human nature.

The fraud was uncovered by the same individuals who initially accepted the Piltdown fake. Tests, unavailable at discovery, confirmed that the skull and jaw were not connected. Indeed, the jaw was of a modern orang stained and shaped to look old and fossilized. The maker of the Piltdown bones knew enough about anatomy to fool some pretty good men. But these same men wanted desperately to believe what the bones said because believing fed their delusions. In the end, the situation was corrected, and anthropology emerged a bit more skeptical and much wiser after dealing with the Piltdown skull.

It is not always so. At about the same time Piltdown was playing out, another group of anthropologists was busy manufacturing and disseminating a concept of human nature that, to this day, haunts us.

Our story starts with Charles Darwin. In 1858, at a meeting of the Linnean Society, the first formal account of descent with modification and natural selection was presented. Much of the science community was actually relieved. Creationism, the then official story of man's appearance, was getting creaky with contradictions. The time frame was way too short; the phylogenic relationships of plant to plant and animal to animal looked suspiciously well organized to be entirely random. And there was no way that a single flood could account for the geological column. So an explanation that fell more in line with observation was pleasantly received in the scientific community.

From the fruit of Darwin's labors, many conclusions were drawn—most realistic and reasonable. For example, all life is related, a point pretty much validated by DNA; organisms vary, another point pretty much validated by both examination and breeding. Many formerly unexplainable observations came into focus, and one contemporary is supposed to have said upon hearing of the theory, "How stupid of me not to have thought of that." Evolutionary theory explained many things, and as a result, its implications were applied to every facet of life, sometimes with less than desirable results.

It quickly became obvious that substantial inherited relationships existed in both plant and animal life. Species could be assigned to a genus of related species; a genus could be assigned to a family, a family to an order, and so on. Various characteristics of species made this assignment possible; relationships were established, and patterns emerged. Then someone got the bright idea that if physical characteristics were inherited, so also might behavior be an inherited characteristic.

At one time, it was assumed that all or most of animal behavior was coded at conception and that the majority of life-forms behaved as a direct consequence of chemical processes in the brain over which the animal had little or no control. Recent investigation disputes this. But there is little doubt that a good part of animal behavior is hardwired in the sense that both action and reaction are strongly influenced by instinct. That is not to say that animals are automatons, but rather, over eons, the response to many stimuli have come to be automatic in the same sense as breathing or the heart beating, something the animal does without thinking. Other responses to stimuli may require a thought process of one kind or another. So animals have both voluntary—and involuntary—behavior potential as mechanisms of survival.

Perhaps, since man was an animal, we, too, were hardwired in some areas of our brain, making our behavior to some extent something over which we, as individuals, had no control. Perhaps, since we were only another kind of animal, we, too, were more or less hardwired in certain areas just as any other species might be. And so grew the great "nurture versus nature" controversy.

It was argued that some individuals were more naturally fit to control and lead and even exploit others because nature had made them that way. Unrestrained and uncontrolled use of labor by management was a product of natural selection. Certain inherent characteristics in some individuals made them superior to the hoi polloi.

And because such characteristics were inherited, it meant that those individuals and their offspring were destined to be rich, and everyone else ought to get used to the idea. This was nature at its very best, rationalizing the exploitation of the masses based on natural selection and justifying the dominance of others as a component of natural law.

Another school of human nature sprang up in the halls of academia to counter this nasty, awful, ill-conceived concept. The human brain at birth, these pontificators pontificated, was blank. It had no instinctive predispositions, no primal urges, no inherent wants or needs; in short, it was a blank slate. A person's disposition, personality, and behavior were all the result of the experience the individual had during their lifetime but especially during their childhood. This was nurture at its finest. If we were all just nurtured correctly, if as children we were given lots of love, if we felt secure, if we were always happy, if we were content, then when we grew up, we would be loving, kind, well-adjusted individuals and be a joy to our mothers.

Nature versus nurture. Are we ravening beasts or gentle lambs? Are we victims of our instincts, or do we control our behavior? Are we truly the epitome, the pinnacle, the glorious finale of evolution, or just another beastie in an ongoing compilation of beasties? The answer is that both nature and nurture impact our lives. The reasons why are fascinating, and we will pursue them in greater detail further on. For the time being, we must deal with the delusions this opposition of opinion has created.

Nurturists did not like the picture the naturalists painted. We were not beasties. We were kind, loving, gentle; and if we weren't, it was because of our upbringing. The nurturists set out to prove this by studying different cultures. The trouble was that every culture studied demonstrated a disturbing trend. A pattern emerged of war, greed, avarice, and deceit. We were, as a species, nasty, brutish, and xenophobic. Even cultures on the fringes of the world, the Eskimo and the !Kung, showed strong signs of animallike behavior—not at all human, not at all nice. Surely, the nurturists reasoned, there must be somewhere, someplace, a group or tribe or culture that exhibited the finer attributes of human nature—a quiet, nonviolent, kind, loving culture. In short, a fitting finish to evolution's plan, a peaceful man. Alas, the nurturists couldn't find one; a dilemma, indeed.

Enter Margaret Mead, a young student of anthropology, and her mentor, Franz Boaz. Both were ardent nurturists. Both were sure their theory of human development depended on upbringing. It was so simple, so refined, so achievable, if only we were nice to one another, if only we just loved one another, if only we were kind to one another. The image was so easy to visualize, so elegant that it must be true. Since we were the culmination of evolution, the concept of our being the best just had to be correct. It just had to be the answer. The question was how to prove it.

It looked bad for the nurturists. Nowhere could there be found a tribe or group that would break the mold—a culture that would show that under the right conditions we were naturally nice, kind animals and as such would display our nice, kind behavioral dispositions. Mead and Boaz decided that if they could find one, just one culture that demonstrated these attributes, then they could prove that it was nurture after all. No matter how many bad patterns in human behavior were cataloged, if one with good patterns was documented, then the reasonable inference would be that all the bad ones were the result of bad nurturing while the lone good culture demonstrated the finer points of natural human behavior.

It was a terrible plan. To prove the gentle nature of man by ignoring all the cultures that flatly contradict your theory and by hook or crook find one that, to a greater or lesser extent, validates your concept of human potential and then say this culture represents reality is hubris on a massive scale. But that is exactly what Mead did.

Where better to find paradise than in paradise. That wonderful world of palm trees, white sand beaches, and deep blue waters of the South Pacific. A place where coconuts and bananas were to be had for the picking, fish for the netting, and where taro literally jumped out of the rich volcanic soil. A place where long, lazy days on the beach were interspersed with hammock resting and pig feasting. If we were found to be nice in this perfect paradise, then Mead and company could argue that this was the true natural state of man. If the South Sea Islanders could be shown to be nice, kind, loving, gentle, and sexually free, then this would demonstrate that the natural state of man was nice. Sexual behavior had a lot to do with Mead's theory. Her contention was that much of the trouble and bad behavior that plagued mankind could be traced back to the sexual repression and that nagging subconscious frustration at the sexual mores of the day. If only we were all free of sexual impediments, then we would become truly nice.

And so Margaret Mead went to American Samoa, talked to two native girls who told her a bunch of lies because Mead's questions embarrassed them, and came to the conclusion that what young girls the world over need was more sex and less parental control. She wrote a book about the few months she spent in paradise. She called it *Coming of Age in Samoa*. The book made her famous, established her as a cultural icon, and created a vast collective delusion that plagues us to this day.

The truth about Mead's stay in Samoa has been more than thoroughly documented by Derek Freeman in *The Fateful Hoaxing of Margaret Mead: A Historical Analysis of Her Samoan Research*. Mead wanted desperately to prove that we were nice, kind, gentle, a fitting conclusion to evolution. She was so sure she was right that she ignored a mountain of evidence, some of which was included in *Coming of Age*, and placed blind faith in a pack of lies. The truth is there in the book and remains there for anyone not under the spell of Mead's delusion. Mead states, "War and cannibalism have long since passed away." But they had not long since passed away. Three hundred years before Mead's arrival, the Samoans were some of the fiercest warriors in the South Pacific. Indeed, all the cultures in the area engaged in warfare on a massive scale, sailing thousands of miles in open canoes to wage war on their fellow Polynesians. Pacification came in the form of superior European weapons and smallpox. So Samoan culture was not so different from other cultures as one might suspect. At one time not too long ago, it was as violent and nasty as any other culture in the world.

Then there was the stratification in Samoan society. There were chiefs, high chiefs, talking chiefs, and a whole raft of hereditary social positions that determined at birth an individual's future status. Hardly an egalitarian society.

But it was the sex thing that Mead should have picked up on. Much of what we think about sex is delusional. But in Margaret Mead's case, it was delusion with a capital D. As mentioned, Samoan society is highly stratified. Individuals in the higher ranks have certain duties and responsibilities imposed as part of their status. One of the most important positions among the females of the chiefly class was that of ceremonial virgin. Ceremonial virgins were to be just that, virgins. Mead knew this. She had accepted a title of ceremonial virgin even though she was not—being a married woman. This fact was kept from the Samoans. Rather than accepting the position of ceremonial virgin, Mead should have considered the contradiction objectively. Sex was okay for all young unmarried Samoans. Sex was good for all young unmarried Samoans. Sex was accepted among all young unmarried Samoans, except for those at the very top of the social order. Mead never bothers to explain the logic behind this dichotomy. Lower-class girls were allowed to have all the fun, the excitement, the pleasure, while high-class girls had to keep their legs crossed. And not just in theory either, ceremonial virgins were subject to a very public and embarrassing rite whereby their purity was verified. Woe be unto those who did not display blood. They could be killed.

So here we have a culture that esteems purity among its females while allowing these same females to engage in sexual activity anytime, anyplace, and in any way. Mead does not find this unusual or odd or even a minor impediment to her theory of more sex and less parental control. Nurture is her bag, and she will substantiate her contention that upbringing is all, no matter what the facts say. Mead and Boaz fostered a delusion, a delusion on a grand scale, all to validate their concept of the nature of man.

Okay, you might say, so she fudged a bit. She stretched the truth. So what? Mead's machinations were directly responsible for the flowering of the counterculture in the fifties and sixties. Make love, not war; peace; and the pipe are all the results of impressionable young reading *Coming of Age in Samoa*, liking what they read, believing it, and living it. A laid-back lifestyle, tune in and drop out, be cool—just like the Samoans. Many a parent decided, on the basis of Mead's book, to allow their offspring a great deal of latitude in lifestyle choice with the hope that less parental control would produce better results.

Alas, many were doomed to disappointment. Some were pleasantly surprised, and the bulk of the fifties and sixties teenagers grew up to be reasonably well-adjusted adults in spite of their upbringing. The problem is that much of Mead's philosophy is still entrenched in academia. We want to believe we are good, kind, rational, loving, and a fitting culmination to evolution. Mead resonates because she feeds that desire. We like what she says. We revel in the implications of more sex and less parental control. And so most of our college humanities students are steeped in Mead and drunk on delusion. For most, experience corrects the delusion. But for a fanatic few, the rose-colored glasses become blinders. The result is academics totally out of touch with reality and more than willing to accept the most outrageous claims with respect to human behavior as long as they are positive. Just once, it would be satisfying to see man painted blacker than he really is, but I doubt it will ever happen.

Mead maintained that more sex and less parental control would produce better-adjusted females. This would be the first step in bringing about a better world. If only we were freed from our stuffy Victorian sexual attitudes, then we could become truly happy, carefree children of nature. Mead promoted promiscuity, believing that such behavior would result in well-adjusted adults. It never occurred to her that we might, in fact, be naturally promiscuous.

Our belief that the human animal is monogamous is a delusion. Our articulated preference is for love requited, a pair-bond, a perfect union. Sounds good, looks better, and it's wrong. A picture is painted of the noble savage and his beautiful, if somewhat pliant, mate walking hand in hand into the sunset in anticipation of a lifetime of huggies. It doesn't work. Or we are told that the male cleaves to his female to ensure paternity of progeny. It's a bunch of bunk.

No social primate engages in pair-bonding within the social group. And that's just for starters. Social primates cannot afford to pair-bond. There are almost always more females than males in most troops. Monogamy would mean that some females would not have mates in a pair-bonding social unit. As a result, some females would not breed, and the troop would be the worse for it.

Compare this to a group of animals that follow the practice of all females breeding with all males. The all-breeding troop would ensure that all females were contributing to the gene pool. Such a group would have more offspring than the pair-bonding bunch. And that bit of leverage would be enough to ensure that group sex survived.

However, our sexual evolution was not without its problems, especially for the female. As we shall see, when we evolved our bipedal stance, the female half of the sexual equation lost the ability to visually display their sexual availability. Scent clues were also reduced to a point where they were not reliable. So what's a poor girl to do? She knows she is ready, her body tells her so, but old caveman over there is busy sharpening his spear or carving his club or simply sleeping off his latest greatest meal.

She could throw a rock, but that would make him mad. She could explain the situation to him, but hominid vocabulary is not that sophisticated. So she does the only thing any self-respecting hot-to-trot female would do: she goes over to the inattentive male and sticks her ass in his face. This has been the standard classic signal for hundreds of thousands of years because human females had no other way to signal sexual receptivity.

Once one male commenced, all the males got in line and paid their obligatory respects. Our lustful lass, momentarily satisfied, repaired to the fire to bask in the attention afforded her by the other females. As we shall see, the sex act itself was not that common in our past. Probably no more than five or six times a year did the clan experience a female coming into heat, perhaps more often if infant mortality was high. But sex, while important from a procreative standpoint, was not the be-all and end-all that it is today. In our hunter-gatherer past, sex played its part as one activity among many. It is entirely possible that our fine, upstanding male ancestor got as much pleasure from killing a bull elephant as he did from coupling. A different kind of pleasure, but pleasure nonetheless. Prehistoric cultures were simple, limited, unembarrassed, and logical. Not so today.

Today we claim monogamy. We like to think of ourselves as pair-bonding—man and wife strolling through life hand in hand, plucking, and gathering, blissfully unaware of the animal ugliness around us. We have seen one of the origins of this delusion. There are more. It is now time to present the evidence for our promiscuity.

Let us start with the ladies. Female sexual physiology is, to say the least, strange for a pair-bonding species. Females take much longer than males to achieve orgasm. In addition, the female is capable of enjoying a number of orgasms in a row, curious sexual behavior for a monogamous female. If, on the other hand, we are polygamous, then the explanation is simple. Several partners would be necessary to arouse the female. But once she was aroused, she could experience a number of blissful bangs. Evolution would select for just such a female because the evolutionary goal is to ensure conception. Her sexual disposition would meld nicely with the evolutionary edict to be fruitful and multiply.

Males, on the other hand, can only ejaculate once and then need a bit of rest, just the amount of time needed to go to the end of the line. However, our lustful males do offer us evidence of polygamy. Penis size among Homo sapiens has been characterized as among the largest of any animal our size and even for some animals much larger. This is puzzling if we are monogamous, certainly not something selected for. Both large, medium, and even small penises would have impregnated our spouses. Well, perhaps we can postulate that a larger device would be more likely to satisfy our partner and so was selected for. But no, female satisfaction in a monogamous relationship is a nonstarter. She would have been impregnated, satisfied or not.

However, if we were polygamous, then tool size assumes new importance. Sperm is deposited in the vagina. From there, it must swim through the uterus to one of the fallopian tubes where it will fertilize the egg. This is a long, long journey, sperm-wise. And any variation that shortens the trip would be selected. Sperm inserted further inside the vagina will have less distance to travel and a better chance of fertilizing the egg. Larger, longer penises will be selected.

There is yet a third indication of our polygamous past, and that is the present. A recent survey of cultures around the world showed that 85 percent of them were polygamous in that males had more than one female at the same time. A few cultures favored the female and permitted her to have more than one male partner. A small fraction were nominally monogamous. But most, in reality, were cultures where males and females had multiple partners. Having said this, the study concluded that we were monogamous, albeit, imperfectly so.

Monogamy is a delusion. Its roots lie in our recent past and are marginally maintained by those in love with the concept of two lovers strolling through life hand in hand, depending on their feelings for each other to get them over the rough spots. A study of cultures show that, with rare exceptions, the concept of marriage is male dominated. It is the males who decide the relative position of females, and the males almost always place the women in a second-class position.

Monogamy is a delusion. It is, however, a delusion that fits in very well with our perception of self. We see the pair-bond as a desirable state of human behavior. Our cultures, which are male driven, see promiscuity as bad, deviant, and not to be tolerated. As we shall see, the males have an ulterior motive for such distinctions.

We cannot begin to fix the problems of the human animal until we understand the human animal. Monogamy is a problem. Delusion is a big problem. We need to know and understand this if we are to understand man.

Margaret Mead wanted desperately for us to be nice. We were the very culmination of evolution and, as such, just had to be the best. This concept of the perfection of the human animal pervades the social sciences. We are nice, we are kind, we are altruistic. We just have to be because we are the finest creation of evolution.

Mead was wrong. The Samoans were and are just like everyone else. Unfortunately, her message resonates. People like to read nice things about themselves. They like to think highly of themselves. They like to think that we have the ear of God, and that we are special.

As we have seen, even the hard sciences such as archeology or anthropology can be and are influenced by man's delusions. Fortunately, the hard sciences are, to a great extent, self-correcting. Soft sciences, on the other hand, can live their delusions long past a point where reality should kick in. Opinion matters, and self-deception survives where objective considerations should apply. But it is with the delusion of religion that we really demonstrate the madness of man.

Religion is a delusion. It is a scam upon humans by humans. Nothing displays this better than that patently false absurdity called creationism. This is our next delusion.

CHAPTER 2

An excellent case can be made for the contention that our delusional predispositions are genetic in origin. There may be selective evolutionary value to believing in the supernatural. It is a matter of confidence or lack thereof. The concept no longer applies as natural selection today has little, if any, significant impact on the human animal. For the most part, we all survive and we all reproduce. But at one time, when natural selection was selecting and when descent with modification was modifying, believing could have been beneficial.

We lived in small groups, we defended territories, and we raided other territories to maintain territorial integrity for females and just for the fun of it. We had limited weapons, and each individual had just about the same potential for doing damage to an opponent. Stones, flint blades, and pointy sticks were simple but effective tools against game and against each other. Our individual strength, too, was almost even. There were no ninety-eight-pound weaklings. So in strength and arms, we were all pretty much equal.

It fell to intelligence to make the difference. More intelligent individuals could use guile and trickery to overcome an adversary. This "modification" was thus selected for.

But an additional potential for winning was confidence. If an individual warrior imagined a supernatural being and further imagined that this supernatural being was on his side, then maybe, just maybe, he could overcome an equally intelligent, equally strong, equally well-armed opponent. The very thought that his supernatural being would not let him down may have been, just often enough, the difference between winning and losing, and thus, the concept of a supernatural was selected for.

It would not necessarily have been a predisposition to believe in God that prevailed but rather a genetic predisposition to believe very passionately in any kind of delusion. Natural selection is not discreet. If a belief in a greater power is brought about by selecting a brain capable of believing almost anything supernatural, so be it. Nature is sloppy. Nature doesn't care. And the result is a brain capable of believing something, anything, without a scrap of proof, believing without the slightest evidence and a total lack of substantiation. As we shall see, in some individuals, nature has overdone the delusion, and these poor souls are capable of clutching the most outlandish, the most impossible beliefs to their bosoms. In a kind of cockeyed way, it makes sense. What the individual is, in essence, demonstrating is that the more outrageous the claim, the more important it is to believe. After all, anyone can believe small delusions. It takes a truly devoted believer to believe in the unbelievable. The more unbelievable these delusions are, the more likely they are to be believed. Perhaps this is why we have so many creationists.

If this is true and a genetic predisposition favoring delusion is deeply ingrained in our brain, it will be almost impossible to erase the predisposition. We should, however, understand the source of our delusions.

We seem to be unusually prone toward impossible beliefs, and the more impossible, the better. If anthropology is to be taken at its word, all cultures believe in the supernatural, the reality of gods and demons, the ability to predict and influence future events through prayer, sacrifice, or simply believing strongly enough. There also seems to be a group of individuals in place who happen to be able to help, for a price. Always for a price. Witch doctors, shamans, priests, popes—whatever they are called—will help, but only if we do as they say. We must defer to them, respect them, hold them in high esteem, and, of course, support them in a manner in which they would like to become accustomed. Creationists fall into this category.

Creationists say that evolution is an impossibility. That we are too perfect, too special, too important to have evolved by some blind chance sort of process. We are unique, we are different, and we are the apex of life. Creationists want desperately to be special. One cannot be special if one is a descendant from an ape. Such a common ancestry from such a brutish beast just will not do. We are special, we are unique, we are different; and so we just must have a pedigree in line with our importance. To be made from whole cloth, that is the past for us. It does, after all, demonstrate our unequaled ancestry. Surely some special guidance was needed to arrive at us. Some divinely driven plan whereby the perfect being, the image of God, was produced. Evolution, they say, is the equivalent of a tornado going through a junkyard and by some chance producing a 747 jumbo jet.

I find the analogy of the Boeing 747 amusing. Evolution does not sweep through a junkyard and "make" a 747. Evolution uses two main methods of "making" different forms of life. Notice I did not say making life. Evolution is not about the beginning of life. It is about the way the different forms of life got to be the way we see them.

Back to our 747. Suppose we apply the principles of evolution to this airplane. The 747 is too big for asexual reproduction, but perhaps sexual reproduction offers more fertile ground for speculation. But questions immediately arise. Is a 747 male or female? Does it copulate? If so, where and with whom and with what? Is each type of plane a species, or are all jets of the same species but different races? What about propeller-driven airplanes? Is it possible for a prop plane to mate with a jet airplane? And what would the offspring be? Prop jets perhaps?

How would one breed a plane? In the air, on the ground, or in the privacy of a very large hangar? And how would they breed? A male bird will mount the female and, from what appears to be a very awkward position, rubs his cloaca against hers. It may not be elegant, but it gets the job done. Would a 747 mount a 737? And what would they rub together? Is a puzzlement.

It gets more so. Assuming a successful copulation, even more questions arise. Where does the "female" plane carry the unborn, in the baggage compartment perhaps? What is the gestation period? Would a pregnant plane fly? Should a pregnant plane fly? Would the female plane lay eggs or motorized hang gliders or baby 747-737 hybrids? (Since we do not have hybrid 747-737 planes, I suppose it would have to be one or the other.) What would the newborn plane be fed? Would it reside in a nest or in a hangar?

Would both parents care for their offspring, or is our 747 male a philanderer, knocking up as many females as possible between fights with other 747s for dominance?

It is just too much. A 747 in a junkyard cannot be compared to life, no matter how hard the wind blows. The 747 analogy is simply spurious. Planes are inanimate; life is animate. People, plants, even bacteria are made with a totally different process. The 747 analogy is only one of many creationist delusions.

We are evolved animals; our life history is that of an evolved animal. What happened in our past impacts us today. The way we behave, the way we live, the things we do are all colored by the thousands and thousands of generations that precede us. If we are to truly know ourselves, we will have to know and understand our past. Creationists claim we have no past, and that is delusion.

The nuts and bolts of this patently false delusion are as follows: All life was created at one time by fiat. After a period of time, much of this life was destroyed in a flood. And only a small portion of the total biomass survived by being carried about in a boat. It was that same flood that rearranged the landscape and produced the world as we see it today.

Is this a possibility? Can anyone really believe that the whole world was, at one time, completely covered by water? I think not. When one looks at the evidence, one is forced—if one is delusion free—to consider creation a real big delusion.

I am going to present the case for three of the aspects of the evolution/creation controversy. I am going to lay out the line according to creationist dogma and then show why evolution is the better answer to what we see.

We are going to look at the geologic column. The geologic column actually exists. It is not a theory or an idea or a concept. It is there, in place, and anyone can go out and look at it, study it, and draw conclusions from it. One of the most common observations made of the geologic column is that plant and animal remains are found in a very consistent order. The fossil record and the geologic column are always the same. We do not see any mixing of the fossils. The plant and animal fossils are separated into layers with the same fossils always in the same layers.

In lower layers of rock, more primitive fossils are found. In upper layers of rock, more complex fossils are found. This is the expectation of evolution.

So how does creationism handle this set of facts? Well, their explanation is twofold. The first, during the flood, the smarter animals were able to run farther up the hills before being overtaken by the rising water. As a result, their bones are found higher on a hill, i.e., in newer rocks.

The idea will not wash. The explanation fails utterly to take into account what happens to animals that drown. In the first place, most sink regardless of where they were on a hill. But very rapidly, bacteria in the carcass begin to give off gasses as a result of the decomposition of the animal. This gas causes the dead animals to become buoyant and to float. So we would have millions upon millions of dead animals of all shapes and sizes in great whacking windrows of bodies wafted hither and yon and all mixed up. When the flood comes, animals do not die and stay where they are at the time of death. The idea that smarter animals managed to climb higher on the sides of mountains and, thus, show up higher on the geologic column is bogus. Creationists say absolutely nothing about trees which—as any biologist will tell you—cannot run uphill.

Their second explanation is that the flood separated and sorted all the fossils in the geologic column, what we see today in the rocks. People actually believe this or, at least, refuse to look at the issue seriously, and that is why one of our most absurd delusions exists.

Creationism is utterly without foundation and is unbelievable. Evolution is both predicted and confirmed by the fossil record, and this can be easily demonstrated.

Let us make a few comparisons. Creationists say that all forms of life that have ever lived on Earth coexisted at one time. All fish, reptiles, dinosaurs, and mammals as well as all kinds of trees, bushes, plants, and grasses (remember the grasses) were living, procreating, and dying in the same habitat at the same time. Evolution says no. Different animals and plants existed at different times. This is why we do not see any tyrannosaurus with elephant tusks through their ribs or modern mammal tracks alongside dinosaur tracks (more on this later), or placoderm fish in the same marine beds as teleost fish. The geologic column screams evolution. Creationists just scream. Who is right? Let us see.

To do this, some assumptions must be made with regard to creationists' claims. But the assumptions are only three and reasonable at that, for they are based on creationists' statements.

First assumption: As we have seen, creationists maintain that all animals and plants that ever lived were, at one time, in the same habitat at the same time. If this were true, then we can assume all these animals and plants also died in that selfsame habitat. The only alternative would be that before the flood, no animal or plant died, and even creationists do not make that claim. If plants and animals died together, then the second assumption would be that in places, their remains would have been buried together—dinosaurs with mammals, Carboniferous plants with Triassic plants with present-day plants, present-day birds with pterosaurs (flying reptiles), insects of the past and present as well as a vast mix of marine mammals, reptiles, trilobites, and other ocean and freshwater life-forms. All these mixed together in the geologic column for several thousand years. Assumption number three is that during this period, some of the remains of these animals and plants were fossilized, leaving us a clear record of this mixed and matched layer in the geologic column.

These three simple assumptions are reasonable, undisputed, and fatal to creationism. The geologic column has been gone over literally with a fine-tooth comb by experts, and absolutely no trace of this assemblage has been found. Not a single bone, not one tooth, no mixed-up plants or insects or fish. Nothing, nada, zip, zero.

How can this be? Well, we are told this flood came along, and it separated and sorted all the animal and plant fossils that were entombed in the geologic column, and placed them in the same position that they would occupy if evolution had occurred. Further, all the plants and animals that were alive at the time of the flood were also deposited in layers to make it look like evolution had taken place. We are not informed as to why this had to be. But creationists even claim to have a formula to prove this extraordinary layering. But it is long and complex, and so is not published. Just how long and complex such a formula would be is about to be revealed.

Let us start with the ocean, for that is where evolution started. First, there are the bones, all fossilized, all in the geologic column in exactly the right place to indicate that evolution had taken place. In the oldest layers (the ones on the bottom) are found simple marine organisms. Among these are primitive jellyfish and sea pens. Then comes the trilobites in many forms and dimensions—shallow water, deep water, blind, and with multiple eyes—all in Cambrian through Permian layers. Then we see evidence of the placoderms, vertebrate killers in a multitude of shapes and sizes. Then they disappear, and soon marine mammals take over as if evolution had actually taken place.

There is in Canada in the western Rockies high up above the timberline a geological formation known as the Burgess Shale. The fossils in the Burgess Shale are unique. They are Cambrian, and they really put a strain on the creationist explanation for separating and sorting of fossils. The Burgess Shale fossils are marine in nature, and all the animals and plants found in the layers of rock appear to have been suddenly covered by a blanket of sediment. Perhaps a series of underwater landslides or maybe a severe storm caused vast clouds of sediment to kill, cover, and, at the same time, preserve *Pikaia*, *Canadaspis*, *Waptia*, and *Wiwaxia* as well as a host of other marine plant and animal remains.

These fossils are preserved in such exquisite detail to be almost unique. Both soft parts and minute details are preserved by the very fine medium in which the animals and plants were entombed. As a result, scientists are able to study not only the hard parts that are normally fossilized but also the body outline and even details of internal structure.

What is peculiar about this assemblage is that although the Burgess Shale contains hundreds of species of plants and animals, there are none in the shale that are alive today. Not one fish scale, not one fish bone, nothing, not even a single shark tooth (shark skeletons are cartilaginous and are not often preserved; most prehistoric sharks are known only from teeth). There is not a single modern marine animal or plant fossil in the Burgess Shale.

The only explanation the creationists can offer is a formula; but a formula to describe the process of separation and sorting by the great fountains of the deep to remove every single modern plant and animal, every fish scale and every shark tooth, while leaving the flora and fauna of the Cambrian sea untouched must be truly formidable. Many lines of x's and o's, w's, y's, pluses and minuses, and square roots to the fifth power over x. Surely such a formula would take up many pages, perhaps even a whole chapter in the creationist formula book.

The problem with this explanation is that much of the fossil evidence in the Burgess Shale consists of discolored shale grains in the shape of the fossilized organism. At death and entombment, the organic material in the specimen changed the hue of the shale. This was not some bone or scale or tooth that retained its shape and outline as it fossilized. If creationist explanations are to be taken seriously, then the only way the separating and sorting could have taken place was for each individual shale grain to be held in suspension while all other fossils were removed and then each grain returned to the rock face in just exactly the right position to reveal the outline of the organism as if no disturbance had ever occurred. This would have had to happen for each and every one of the hundreds of thousands of fossils that lay buried in the Burgess Shale.

It doesn't stop there. Recently in China, other shale beds have been found. They are located in Cambrian rocks, the same as the Burgess Shale. They contain many of the same plants and animals as the Burgess Shale. And again, not one single fish scale, not one bone, not one tooth is to be found. Truly it would take an impressive formula to explain the Burgess Shale and the Cambrian world of *Wiwaxia.*

The formula becomes even more complex, and the reason is fishy. Marine deposits pose a particularly challenging problem for creationists. What impact did a flood of water have on water? Fish don't drown, and yet we are told that all life that breathed, with the exception of those on the ark, were drowned. So fish shouldn't have been affected (unless you count breathing water, something creationists do not do), but they were. At least the fossil record shows a separating and sorting that would take at least a chapter or two of creationist formula to explain.

There are the Placodermi, diverse and Devonian bottom-feeders, open-water breeders, fast and slow, big and small armor-plated-jawed fishes. Lots of species, a number of genera, and more than twenty families. A great many successful fish both fresh and salt-water habituated living in shallow lakes, rivers, estuaries, and open ocean. And they all disappeared in the end of the Devonian about 360 million years ago.

Then there are the Osteostraci and the Heterostraci, some of the earliest of fishes according to evolution. They, too, do not occur in any geological strata above that of the Devonian. The Osteostraci and the Heterostraci have one thing going for them though. They originated in Ordovician marine environments even before the Placodermi. Then there are the Symmoriidae, which died out in the Permian; Redfieldiiformes, which lasted until the Jurassic; and then the Semionotidae, which stuck around until the Cenozoic. Thousands of species, hundreds of genera, and a host of families all separated and sorted, all assigned to layers like files in a cabinet, all living in water and yet separated and sorted by the very same water to give the impression of evolution. Not one fish out of place, not one bone, not one tooth, nothing. It will take a whole raft of creationist formula to explain all that fishy stuff. But there are still more complications to come.

It also turns out that modern fishes are only found in the upper layers of the marine geologic column. Among the teleosts, which include herring, trout, salmon, flying fish, and guppies, none are found in Triassic marine deposits and only one in Jurassic deposits. Almost all are found in middle to late Cretaceous and some in only Cenozoic deposits exactly as one would expect if evolution took place.

Then there is the tale of the trilobite, the *Elasmosaurus*, and the whale. Trilobites first appeared in the Cambrian and disappeared at the end of the Permian. There is not a single trilobite fossil in Triassic or later marine beds. And there were billions and billions of trilobites. In Ordovician shales around Shropshire, England, fossilized trilobites literally make up most of the shale deposits. Cambrian rocks in the vicinity of Shandong, China, contain millions of trilobite parts. But not a single trilobite passed the Permian. Trilobites, dead and alive, managed to ensconce themselves in the fossil record in just such a way as to give the impression of evolution.

Elasmosaurus was a marine reptile that is found only in late Cretaceous marine deposits. A member of the superfamily Plesiosauroidea, *Elasmosaurus* was the last of its kind. Its claim to fame is seventy-one neck bones making its neck at twenty-six feet, longer than its body at twenty feet. Long neck and all, it disappeared the same time the dinosaurs did, perhaps for the same reasons. It left its mark on the Cretaceous but is found no higher in the geologic column.

Whales are an entirely different matter. Whales exist today. Unfortunately for the creationists, they do not appear in any geologic layers older than the Eocene. Whales in the fossil record begin with *Pakicetus*, an animal that lived on land and only shows a few characteristics that point to its being ancestral to Cetacea. By the late Eocene, whales had fully taken to the sea, and one of the best known is *Basilosaurus*. Sixty-five to eighty feet long, *Basilosaurus* was the largest mammal around, and it still had the very small hind limbs of its *Pakicetus* ancestor, limbs that were totally useless but contained all the limb bones of fully developed legs. This condition makes evolutionary sense while at the same time exposes creationism as nonsense.

When land animals return to the water, this new environment puts different selective pressures on the animals. Otters and beavers have webbed feet. At one time, their ancestors did not. Even the slightest bit of webbing between toes would have been marginally advantageous, giving the bearer a slight benefit over others of its species. The marginally better-webbed animals would, in the long run, be more likely to survive and pass on their webbed-ness. So beavers and otters today sport webbed feet.

It works the other way too. Appendages not used tend to atrophy. As *Basilosaurus* returned to the sea, its back legs became less and less necessary while its front legs were slowly turning into flippers. Useless hind limbs (really tiny ones) speak volumes from an evolution standpoint. But who would create a whale with legs it could not use?

Then there are the modern whales, the mighty blue, one of the largest animals ever to have existed on Earth. We also have the humpback, the right, beluga, sperm, and gray whales. All alive today, all marine, all over the oceans, and their remains are found only in the topmost layers of the geologic column. No whales in the Cambrian, none in the Permian, absolutely none in the Jurassic of dinosaur fame. Nada. Zip. Zero. It's enough to make a creationist cry.

These are—from *Wiwaxia* of the Burgess Shale to all the fishes through the *Elasmosaurus* and *Basilosaurus* to our modern whales—all marine creatures. If creationists are correct, they all lived together, died together, sank to the bottom of the ocean together, and were entombed together. Then this flood of water—in the water—separated and sorted all the bones and all the teeth and all the body parts.

All of them—mind you, not 99 percent, not even 99.9 percent, but all of them—to give just the impression that evolution had taken place. As for the creationist formula that explains all this, by now it occupies several rather large volumes, but unfortunately our exposure of creationists' patently false absurdities still has a ways to go.

The evidence of evolution is as clear on land as it is in the sea. The creationist might be able to claim that each bone, each tooth, each skeleton was separated and sorted by the great fountains of the deep into strata that just happens to mirror evolution. But some things are more difficult to separate and sort.

When an animal walks on wet or even damp ground, it will leave an impression of its passage. It will leave a track. Tracks can be long or short, one print or one hundred; some stretch for thousands of feet, and they tell interesting stories.

One such story demonstrates both the truth of evolution and a creationist delusion. The state is Texas; the city, Glenrose; and the site, the Paluxy River. In 1938, a man by the name of Roland T. Bird discovered dinosaur tracks in the bed of the Paluxy River. Many of the tracks were excavated and now reside in the American Museum. Some of the tracks looked vaguely human, but only if you closed one eye and squinted through the other.

Still, creationists jumped on the idea of dinosaurs and man coexisting, the truth of which would validate their model of Earth history. Some individuals, not satisfied with the very vague resemblance, carved little toes into the prints. The better to confirm their claims.

The creationists' caterwauling over the Paluxy tracks led some respected scientists to go to Glenrose and see if there was any truth to these claims. There was no proof. Excluding the tracks that sported little piggies, which even creationists acknowledge were bogus, the remaining tracks were all shown to be dinosaurian in nature. The evidence for reptilian origin of the tracks was overwhelming and irrefutable. Man and dinosaur did not walk together in Texas.

There was also something else, something nobody, not even the respected scientists, noticed, or at least they did not comment on. Not only were there no man tracks in the Glenrose formation, but also there were no other mammal tracks at all. No buffalo, no deer, no elk, no bear, no mountain lion, nothing except dinosaur tracks and, in places, a few bird tracks. And the Glenrose formation is huge, covering thousands of square miles of Texas. Not a single mammal track. Not a mammoth, not a camel, not a horse, not a lion, not a *Megatherium*, nothing.

The southern buffalo herd was one of the largest congregations of animals in the world, numbering in the millions of animals, and its main stamping ground was Texas. Surely one buffalo would have left at least one line of tracks in the Glenrose substrate out of all those millions and millions of animals. Surely more than one deer must have come down to the river for a drink. More than one elk must have mingled with the dinosaurs. Bears must have explored the river bottoms. Packs of porkers must have sought refuge among the dinosaurian browsers. But there is nothing, not a single mammal track.

It gets worse. There has not been a single instance where even one modern mammal track has been found in the same substrate as dinosaur tracks. Not in Texas, not in the western United States or China or Korea or Africa or Europe or South America. Not a single track, not a single sign of mammalian presence.

But there is still more; we do have tracks of ancient humans in Africa. At Laetoli in Tanzania, there are the footprints of a trio of tracks which are bipedal and humanlike. Not human because the tracks are 3.7 million years old. These tracks were made by *Australopithecus afarensis*.

It begins like this: A volcano erupts, not some monstrous belch but rather a refined burp. A fine fall of ash settles over the plain at the base of the volcano. A light rain falls, wetting the ash. Numerous animals and birds walk through the wet cement-like surface. Among the life-forms imprinting the wet ash are three tracks that have the classic bipedal heel-and-toe, ball-and-arch print of protohumans. The ash then dries, preserving the track in the by-now-stone-hard ash. Then the volcano burps again and covers the tracks with another ashfall. Over time, the ash layers build up eventually to be worn down, resulting today in indisputable evidence that an ape, probably lately aquatic, walked the plains of Africa on two legs.

But there is still more. Along with the protohuman tracks, we have tracks of hyena, antelope, buffalo, elephant, guinea fowl, rhinoceros, giraffe, and rabbit. But not a single dinosaur track. In fact, wherever we find large mammal tracks, we do not find dinosaur tracks, not ever, never. While at the same time when we do find dinosaur tracks, we never find large mammal tracks. If creationism is correct, all these animals mingled and mangled one another in the very recent past.

How is this to be explained? Evolution says that these animals lived at different times. We would not expect large mammal tracks and dinosaur tracks to be found in the same strata. The creationist's explanation is as follows: "The great fountains of the deep" separated and sorted the tracks. All the mammal tracks were removed from the sites where dinosaur tracks were, and all the dinosaur tracks were removed from the mammal trackways. Not only that, the "great fountains of the deep" filled in the holes left by the removed tracks to give the impression that they were never made in the first place. We are not informed as to why such an action was necessary. But there is really no explanation required as far as creationists are concerned. I don't know if the creationist's separating-and-sorting formula covers this, but if it does, it must be very long, occupying many volumes—perhaps three or four large libraries' worth.

None of what creationists assert to be reality is real. And this can be demonstrated simply by viewing creationist claims with a little intelligence and skepticism. Each time creationist arguments are approached in an unbiased, informed manner, they are completely refuted. And this is not just an academic disagreement. It has important ramifications with respect to human behavior, as we shall see. We, like all life on Earth, have evolved from other life-forms. And today we carry some of the baggage of that evolutionary relationship with us. If we are to truly understand human behavior, it can only be done from a base of reality, not from a flight of faith.

One more proof in the geologic column, that evolution is reality and creation is myth, and then I am going to stop. Creationists say that the geologic column is the way it is because a great flood separated and sorted all signs of life to give the impression of evolution. This is their explanation for what we see in the fossil record. It is hard to visualize a flood that would impact marine deposits in such a way as to give the impression that evolution took place. It is even harder to visualize how a flood could separate and sort animal tracks in such a way as to make it appear evolution took place. It is really impossible to see how a flood could separate and sort pollen grains to validate evolution. But if creation is correct, somehow, someway, it did. Pollen is the male sexual component of plants. Pollen grains are small, very, very small. They are also covered by a sheath called exine, a nearly indestructible coat that helps preserve the grains even after the pollen has lost its ability to fertilize the plant for which it was intended.

Pollen is not just a grain. Each family group of grains is distinctive enough, the palynologists (people who study fossil pollen) can tell which family of plants produced the pollen. Some have bumps in patterns, some have hollows in patterns, some look like ripples on a brain, and, of course, they vary in size. Typically, pollen will range between ten micrometers and two hundred micrometers in diameter. Ten micrometers is equal to 0.00039 inches or about four 10,000ths of an inch. Pollen is very, very small.

One other characteristic of pollen is that there is lots of it. A single pine tree may produce 320 billion pollen grains in a 50-year life span. Multiply 320 billion times the billions of pine trees in the world, and you get some idea of how many pine tree pollen grains there are. And that is only pine trees.

So what we have is trillions upon trillions of pollen grains and spores from billions and billions of plants, including grasses (the grasses are important), produced over time, and because the shell of the pollen grain is so hard, it remains in the soil and sedimentary rock in the geologic column indefinitely.

In the beginning, it was not pollen but spores that were found in the geologic column. Spores, unlike pollen, can set up shop, so to speak, as they are a kind of reproductive source. If spores find fertile ground, they can start to grow and reproduce. They are, however, very small. Larger than pollen, but not by much.

There is, for example, *Cooksonia*. *Cooksonia* was one of the very first of the plant families to exit the sea and invade the land. A simple plant, consisting of stems and sporangia, it had no leaves, flowers, or seeds. It is probable that it did not even have roots, just stems with tiny hairs connecting the plant to the soil. *Cooksonia* did, however, have one characteristic of modern ferns. It had spores. Producing them by the sporangia at the tips of the stems, the plant was nothing more than a spore factory on tiny stilts. *Cooksonia* spores are found only in the same geologic layers as the plants. So if creationists are correct, they must have been somehow separated and sorted in order to appear in the geologic column as they do.

Cooksonia did not last very long though. It came ashore in the Silurian and was extinct by the early Devonian, a victim of more advanced plants. *Cooksonia* spores are found only in Silurian and early Devonian geologic layers. Tiny, tiny spores, too small to even break the surface tension of water, regulated to two consecutive layers of the geologic column.

Or consider the genus *Psilophyton*. It does resemble plants a bit more than *Cooksonia*, having thornlike shoots that resemble primitive leaves. It was bigger, too, reaching heights of fifteen to twenty inches. *Psilophyton* and its distinctive spores are found only in lower to middle Devonian sedimentary layers.

Or perhaps *Lepidodendron*, a tree six feet in diameter and one hundred feet tall. It was common in Europe and North America. It came and went in the Carboniferous. And so its pollen, along with its fossilized plant material, is found only in Carboniferous strata.

Sigillaria offers another example of the reality of the fossil record. This is a tree with spore-bearing cones; it existed from the lower Carboniferous to the lower Permian. Those two geologic layers are the only ones to contain its fossilized remains including spores.

Then we have *Annularia* and *Asterophyllites* restricted to upper Carboniferous and lower Permian strata, *Calamites* from the lower Carboniferous to the lower Permian. There is a large family of seed ferns called pteridosperms, which produced pollen. They evolved first in the Devonian and became extinct in the upper Cretaceous. Plant remains and pollen grains of pteridosperms are found only in Devonian through Cretaceous sediments.

The pollen problem works the other way too. We talked about pine trees and their load of pollen—trillions upon trillions of pollen grains over millions of years. *Pinus* first came upon the geologic scene in lower Cretaceous times. Conifers of all shapes and sizes from a very restricted *Glyptostrobus* to the very common lodgepole and Scotch pine are all producing pollen, and none of it in any geologic layers below the Cretaceous.

And then we have the flowering plants. They have not been with us always, having evolved in the Cretaceous. And sure enough, no flowering plant pollen has ever been found in place in sedimentary layers older than the Cretaceous of 135 million years ago.

But it is the grasses that really cause the problem for creationists. Grasses are everywhere. Vast plains that cannot support trees are wet enough for grass. Millions of square miles of grass—as far as the eye can see—in North and South America, in Europe, in Africa, in Asia, lots and lots of grass. And grasses produce pollen—lots of pollen—because grasses are wind pollinated and must produce inordinate amounts of pollen in order to assure that at least some of it will end up pollinating the grass seed. Over millions of years, grasses have produced uncounted trillions of pollen grains. If ever there was an uncountable number, it would be that number of grass pollen grains produced in the last 135 million years. Seas of grass, waves of grass, prairies of grass, hills and valleys of grass, all producing pollen, all wafting pollen grains onto gentle zephyrs or onto gale-force winds. Pollen beyond comprehension and not a single grain appears in the Carboniferous, Devonian, Silurian, Jurassic, or Cambrian strata. Not one single grain.

Now creationists might be able to convince some easily duped individuals that the "great fountains of the deep" just happened to separate and sort all the bones and teeth in the fossil record to imply evolution. Creationists might even be able to sell the snake oil stupidity that at one time all the animals comingled and left mud tracks all over the place. But the "great fountains of the deep" removed all the mammal tracks from all the dinosaur tracks and all the dinosaur tracks from all the mammal tracks, without leaving any trace of that very nifty sleight of hand. But can anybody really be expected to believe that the "great fountains of the deep" were able to separate and sort each pollen grain, each spore, so that it ended up in exactly the right layer of the geologic column to give the impression of evolution? Pollen grains are so small, so minute, they cannot even break the surface tension of water.

A formula long enough to explain such an occurrence would require enough books to fill every library that exists or has existed in the past ten times over. So we can see just how long the creationist formula for "explaining" the fossil record in the geologic column is. We can also see how silly it is.

Creationism is a delusion. Its attraction lies in the supposition that if we are created, then we are special, and if we want anything, it is to be considered special. This is the great creationist attraction. If, as I believe, we are genetically predisposed to find ways of expressing our specialness, then it will be almost impossible to convince a creationist that they clutch delusion to their bosom. Their very thought processes are such that they reject any argument from reason. Christian creationists just know that they are right. What is just as interesting from a skeptical standpoint is that a great many other religions, vastly different from Christian creation, also think that they are right. They cannot all be right, but they can all be wrong.

There is yet another problem associated with creationist pontificating. Christian creationism likes to postulate a them-or-us confrontation. It is either Christian creation or evolution, they say, and the powers that be should present the arguments for both sides and let the people decide.

But it is not just Christian creation or evolution. Christian creation is just one among many creation myths. Like all creation myths, it suffers from one overriding debilitation. No matter what point is made in reference to the Christian creation myth in an attempt to disprove it, the answer is always the same, "God did it that way." We do not know why he did it that way, but he did. The problem is that the same argument can be made for any creation myth. No matter how absurd, how far-fetched, how ridiculous, the answer is always the same, "God did it that way."

As we have seen, the fossil record is an absolute, total unmitigated disaster for creationists. Everything about the geologic column demonstrates the validity of evolution and the absurdity of creationism. The creationist myth answer to the geologic column is "God did it that way." The creation of light in the process of shining, "God did it that way." A worldwide flood and a boat and eight people all because "God did it that way." But let us see if this is really an adequate answer.

Among the Yao of Mozambique, the creation story is one of decent, kind gods and nasty humans. In the beginning was the Yao god and all the peaceful beasts of the world. Then one day, Chameleon caught a small man and a small woman in a fish trap. He took his catch to Mulungu, the Yao god, and asked him what to do. Mulungu said, "Turn them loose." A big mistake because the man and woman grew and proliferated to a point where they chased Mulungu out of the land, and now he lives in the sky, a place where man cannot reach.

Is the Yao's Mulungu real? What evidence do we have of his reality? None, really. The Yao say that's how it happened, and their assertion cannot be proven or disproven. Just like the Christian creationist myth.

Consider the ancient creation story of Egypt and Atum. One day, Atum was feeling bored and began to play with himself. He began to masturbate and, at his ejaculation, used the semen to make man. One of the more poetical statements of this myth deals with the actual masturbatory act. According to the story, Atum "had union with his clenched hand and joined himself in an embrace with his shadow." Kind of tells it like it is, doesn't it?

He also made everything else we see in the world today. Can it be demonstrated that the Christian creation is true and the Egyptian creation false? The answer has to be no. Any statement made to demonstrate the viability of one myth can also be used to validate the other. Actually the Egyptian myth has a leg up on the Christian creation. The Egyptian god used semen to make man and, in the process, got the concept at least half right. Semen is one of the two necessary ingredients required to make man, the other being the egg. Christian creation has man made from clay, and that, on its very face, is false.

Another creation myth comes from the city of Eridu in Babylon. In it, the god Marduk makes everything from the "spring which was in the sea was a water pipe." Marduk made the cities first, then dirt, then the beasts of the field and the rivers, plants, grass, and trees; and from a liaison with the goddess Aruru, he created man. So while Christian creation has us looking like gods, Babylonian tradition has us actually descended from gods.

The *Rig Veda* 10:129 called "In the Beginning" recounts its own creation myth. But unlike all other such myths, the *Rig Veda* truly tells it like it is. In effect, the *Rig Veda* states that only God knows or, as the hymn ends, "perhaps he does not know." Unlike Christian creations, the Rig Veda says no one can know the mind of God.

There are many creation myths—all different, all more or less unique, all believed by the true believers of that particular myth—and all, in the final analysis, make the same statement. God did it that way. The phrase applies equally and without benefit to any creation myth.

The Cherokees have a fine creation legend in which the animals of this world came into being from another world "beyond the arch." Evidently, it got kind of crowded "beyond the arch," and they needed someplace else to go. Earth came from below the sea and was very soft and wet. The animals could not come down until it dried out. Man was made after the animals, a brother and a sister who multiplied and peopled Earth.

The Cherokee creation legend shares one principal feature with Christian creation myth. It cannot be verified by any means known to man. Even more important, it cannot be falsified. Falsification is important to a theory that wants to be taken seriously. There must be some reasoning that, if applied, would demonstrate the falsifiability of the theory. For instance, if we found dinosaur and mammoth bones mingled in the undisturbed geologic column, then evolutionary theory would be in serious trouble. If buffalo hoofprints were to be found in the same sediments as dinosaur trackways, then evolution would be called into question. If the DNA relationship of life failed utterly to reflect the classification of life-forms, then evolution would be in serious trouble. So evolution has the potential to be falsified, but no creationist will ever admit that there is the slightest possibility of their pet theory turning out to be a patently false absurdity.

The most recent reincarnation of creationism is intelligent design. This is the sophistry whereby it is declared that we are just too special, too important, just too unique to be the result of random processes. We must have been intelligently designed. It is just too obvious to argue.

But when we apply the question of why to intelligent design, the arguments begin to break down. Why would an intelligent designer use the process of evolution to achieve intelligent design? Over 95 percent of life on Earth has gone extinct as a direct result of the so-called intelligent design of this intelligent designer. Does this really sound intelligent? Does this really sound designed? And who, in their right mind, would use meteorites and ice ages and other disasters to wipe out life on a massive scale for some design? For some perceived purpose? Is that really what an intelligent designer would do?

Another why. Why is the intelligent designer designing? For what purpose? To achieve what goal? Designers design for some purpose. Engineers build bridges and buildings to transport and contain. Architects design for a reason. Cars and boats and buses and planes are designed to carry people from place to place. Appliances are designed to make our lives easier. Designers design for a purpose. What is our intelligent designer designing for?

If this entity knows what he (she or it) is doing, if there is some goal to be reached, some end to be achieved, then we are all in trouble. If this designer is an experienced designer and has a finished product in mind (a universal washing machine perhaps), then we have no free will. Whatever we do, whatever we say, however we act, such behavior has been foreseen by the intelligent designer and is part of his vast cosmic intelligent design. Under this thesis, no one can be held responsible for his/her actions as all such actions are planned for by the intelligent designer and are exactly what this entity wants. Or perhaps there is no purpose, no goal, no final result. Perhaps the intelligent designer just started the ball rolling, so to speak, and is now waiting to see the result. But then, that would be impossible to distinguish from blind random evolution.

This is not just argument for the sake of argument. The human animal has a multitude of very real, very serious problems. Many of these problems can be traced back to our evolutionary past, a past that creationists say we do not have. We are animals, evolved animals. We behave, many of us, like animals. At one time, as we shall see, our survival depended on acting like animals.

Creationists claim that we are special. They claim that we have a relationship with God. They claim that we are important, necessary, and absolutely essential to some vast cosmic plan. In order to validate this contention, creationists are forced to believe in some pretty unbelievable things. Such belief is the very crux of faith. The more outlandish and the more improbable the belief system, the better. Anyone can believe in logic and rationale; it takes a real believer to believe in creation.

CHAPTER 3

Delusion is one of the defining characteristics of our species. The phenomenon is more common among the ignorant, gullible, and superstitious. But even intelligent, educated, and supposedly rational people can and do suffer from delusion. Witness the late James Irwin (colonel, USAF retired).

James Irwin was the only person in the world to both walk on the moon and fall off Mount Ararat. His walk on the moon was the culmination of years of training and studying and decision-making. All of which was designed to protect and help James Irwin survive a very hazardous and risky mission. That Irwin was able to carry off this feat is testament to the effectiveness of his ability to absorb knowledge in a systematic and comprehensive way, use that knowledge when necessary, improvise when required, and perform dangerous missions with remarkable success. James Irwin was a very knowledgeable man, a very resourceful man, and—one would think—a very rational man. After all, he had a master's degree in aeronautical engineering from the University of Michigan. His knowledge and ability made it possible for him to walk on the moon. He fell off Mount Ararat because of an April Fools' joke.

Our story begins in Germany in the early part of the twentieth century. On April 1, a German paper, *Kolnishce Illustrierte Zietung*, printed an article about the finding of the remains of Noah's Ark on Mount Ararat. The story was treated as straightforward news and reported as such. The story of this event was next picked up by a Russian publication and printed as factual. The next month's publication of the German paper revealed the truth that the whole article was merely an April Fools' joke. The Russian publication did not print the second article. James Irwin and his companions relied on the Russian article without bothering to verify its veracity. As a result, James Irwin went to Turkey, climbed the mountain, fell off, and came away with cuts, bruises, and broken teeth but with his delusion intact.

There is a simple, reasonable, factual argument that can be made, demonstrating the impossibility of a worldwide flood, and Irwin should have been aware of it. He went to Ararat, he floundered around in the snow—and he should have known.

At fifteen thousand feet or so, water turns to ice. At twenty-nine thousand feet, we have a world of permanent ice caps and glaciers. As the water from the flood rose, it would have encountered air that got colder and colder. It would have started to freeze. At twenty thousand feet or so, the water would have frozen solid, and any additional precipitation would have come down as snow. Every animal and plant on the boat would have frozen to death. Temperatures at twenty-nine thousand feet (the height needed to cover Mount Everest) are thirty to fifty degrees below zero. Noah could not have survived.

Irwin should have known this. He went to Ararat, he saw the snow, he even walked on it, and he experienced the cold; he knew from his time at NASA that outer space was frigid and that the higher you go in the atmosphere, the colder it gets. This is not rocket science. He should have been able to reason out the impossibility of the flood easily, and yet he did not. And the question must be why.

Simple, straightforward questions arise. If there was a worldwide flood, where did all the water come from? Where did all the water go? Where are the signs of the flood? How could a small boat hold all the animals and plants? How could eight people take care of, feed, and clean all those animals? How could many of the animals get from a snow-covered perch—cold and forbidding—to warmer climes? Water, water everywhere and not a drop to drink. But that is just another problem for the ark. It is not just the water and the ark. It is the miles and miles of floating detritus—trees, lumber, bushes, leaves, all kinds of floating stuff but mostly trees. These would form vast rafts as the prevailing winds acted upon them. As the flood went—where?—away, these vast deposits would have come to rest just as the ark supposedly did on the flanks of mountains. Such large deposits of organic material would still be identifiable today if they existed. What happened to all the flood debris? Many questions—simple questions—ignored questions, ignored because when confronted, such questions lead inevitably to doubt, and doubt is fatal.

To doubt is to die and not be born again. To doubt is to forgo eternal life, and as long as we live, eternal life is what we crave, long for, desire, and demand. So anything—no matter how bizarre, how outlandish, how unreasonable—is to be believed, clutched to one's bosom as immutable truths, which will afford salvation.

Irwin's delusions are a perfect example of my contention that the human animal is biased in favor of a genetic predisposition toward delusion. As we have seen, delusion is—I believe—deeply embedded in our genetic makeup. Irwin demonstrates this perfectly. An intelligent, educated, and trained engineer able to wall off portions of his mind and accept, without the slightest question, something that is obviously, factually, a patently false absurdity.

Where religion is concerned, the mind must not, under any circumstances, entertain doubt. The greater the lunacy, the greater the predisposition to believe. A case can even be made for a need for the belief system to be outlandish. The more unbelievable the myth, the easier it is to believe. This demonstrates that the believer is so committed to the myth that nothing can sway that belief, not even simple logic. There can be no doubt.

How outlandish? How bizarre? Well, if we could, we would ask Marshall Applewhite and the cult Heaven's Gate. He managed to convince otherwise intelligent, rational people that a spaceship was hiding behind Comet Hale–Bopp. So, following his delusion, his disciples donned identical sneakers and killed themselves, convinced they would be transported to the spaceship. The followers of Marshall Applewhite paid for his delusion with their lives.

Or consider David Koresh. Koresh liked to play sex games with pubescent girls. Society, as a whole, does not approve of this kind of behavior especially if they cannot join in. So Koresh had to find a way to indulge his passion while circumventing social mores. The result was his cult. By making sexual activity with young girls part of his belief system, he was able to indulge himself without suffering the condemnation of society, especially since he kept the activity restricted to the young females of his followers. The result of this delusion was death for Koresh and his disciples.

Delusion and death on a larger scale was the result of the cult of Jim Jones. Over nine hundred of his followers drank poisoned Kool-Aid because he told them to. Death on a massive scale, and it's all because of dominance and delusion. Jim Jones aspired to have dominance, control, and power over others. He found ignorant, gullible, and superstitious people willing to submit to his blandishments. When the jig was up, he exercised his ultimate power—the power to take someone's life—and all died for Jim Jones's delusion.

As we shall see, delusion can mean death on a massive scale. Jim Jones, as bad as he was, was a piker compared to some. But first, a short diversion. Suppose creationism is correct and could be demonstrated scientifically. Our scenario would go something like this.

The geologic column is only part of the evidence we have with respect to the age of the Earth and of the universe. Present astronomical theory says that our solar system is 4.5 billion years old and the universe, as a whole, is 12 to 15 billion years old. This 12-to-15-billion-year figure comes from observation. We can see that far out into space, and that means that the light from these distant galaxies has been traveling that long.

Creationists do not dispute this but claim that Earth is only about ten thousand years old and that the light from these distant galaxies was created in motion. The reason being that man needed the light from the stars to be for signs and portends. This demonstrates the creationists' contention that man is not only special but also important. God made special provisions just for us.

But think of how neat it would be if that light had not been created, traveling but rather as it would naturally occur. Every few years or so, a new star would appear in the sky as light from that star first reached Earth. In the beginning, there would be myths to explain this phenomenon: the gods were lighting fires in space, or perhaps it was another god being born in the heavens, or because we were so important, a many-eyed god opened yet another eye the better to keep an eye on us.

Then telescopes would be invented, allowing man to see into space. Scientists would calculate the speed of light, and one astronomer one day would have an epiphany. What if what we were seeing was light that had just reached us from a star? If that were true and the distance to that star could be inferred in some manner other than by using the speed of light, one could come very close to calculating the true age of the universe.

The race would be on. First, measurements would be taken using Earth as a baseline. Two telescopes one thousand miles apart would measure the angle between an imaginary baseline running between the two telescopes and the star. Knowing the angle and the length of the baseline would give the star's distance by triangulation. But the thousand-mile baseline would be way too short to make triangulation over any distance accurate. The next step is to measure the angle of the star then measure it again six months later. This uses Earth's orbit around the sun as a baseline, and a fairly accurate one at that. Some stars have been measured using this system with very satisfactory results. Let us say that our imaginary astronomers do this. And the figure comes up to ten thousand light-years away, give or take five hundred light-years.

Not bad, but we can do better. A probe is launched in the plane of Earth's orbit with the sole mission of measuring the angle to a new star as it first appears. Meanwhile, telescopes are set up for the express purpose of timing the arrival of light from a newly discovered star as accurately as possible.

A new star appears. The probe, out beyond Saturn by now, sends back coordinates, and a new figure is calculated as 9,500 plus or minus 250 years. A year passes, a new star, a new set of calculations. It is 9,650 years with a plus or minus of fifty years. Good, but we can do better; more accurate telescopes are built. A probe is sent out in exactly the opposite direction as the first. Now we have a really big baseline and accuracies in the realm of nanoseconds; scientists are closing in on the age of Earth.

A new star turns on. Calculations are done on the latest, fastest, most powerful computer in the world, and the answer is 9,687 years, 48 days, 14 hours, and 26 minutes, plus or minus 3 minutes. Hands are shaken all around, the champagne is uncorked, and the Nobels are passed out. Science has confirmed the age of the universe.

But it is not to be; the light, it comes, it goes, it is everywhere, and it is old. So it must have been created in process. For what reason, we can only guess. For what purpose, we do not know. And of course, it puts creationists in a bit of a pickle because light coming in from the stars is not nine thousand years old. It is billions of years old, and we have only their word for why.

There is yet another Christian delusion that sounds good, does much for our image, and allows us to bask in the rosy glow of satisfied humility. We are, according to Genesis, created in the image of God. That is all fine and wonderful and satisfying. It's a fitting frame for a crowning achievement. It would be good to be a god, but barring that, at least we can look like one. But what, after all, does it mean?

Genesis 1:26 says "in our image, after our likeness," but what does that signify? Does God have a belly button? If he doesn't, then we were not created in his image, none of us. If he does, then that implies a mother of god. If God is great, what does that make his mother? Does God have a mouth, a gut, an anus? Does he eat? Does he defecate? What color is God? Is he brown, black, white, red, or does he have that beautiful western Mediterranean olive tan glow? If he is any of these colors, then the majority of humans were not made in his image. If he is some other color, then none of us are made in his image.

Is God male or female? One or the other and you automatically exclude one half of the population. If God is female, does she have a uterus, vagina, or fallopian tubes? Does she menstruate? Does she have PMS? Are all the floods, hurricanes, and tornadoes simply the signs of a goddess in the throes of the "curse of woman"? If God is a man, does he have nipples on his chest? If he does, why? Does God have a penis? Does he use it for anything besides urination? Does he have testicles, and if he does, why does he have testicles?

It is better not to ask. We are created in the image of God; that is enough. It feels good; it gives us a sense of our own importance. It is also delusion.

Another of the delusions Christians find so seductive about religion involves the fight between God and Satan. It must be deliciously satisfying to know that you are so important that the gods fight over you. The truly contented Christian knows his standing in this war—he is the prize, the goal, and he knows that God and the devil both want his soul. This thought must be wonderfully calming to those unsure of his or her worth, hesitant of their value, or doubtful of their superiority to others. After all, if the gods are fighting over one, it means one of two things—either one is important or the gods are crazy.

Are we really such an important species that the gods spend endless time, energy, and thought on us? Well, let's see. There have been numerous cities and even states destroyed to testify to the importance of man—Sodom and Gomorrah come to mind. Once, according to the Bible, the population of the entire world—along with all the plants and all the animals, with the exception of one family and a favored few beasts—were destroyed. As a species, were we truly so special that everything had to be destroyed just to make a point? Or is this just another myth designed to show that no sacrifice is too great when it comes to man?

Sacrifice is common in the Bible. And it was common in most societies throughout the ages. At the very beginning of the Bible, we have Abraham ready to sacrifice his son Isaac to God. To sacrifice a son, especially a firstborn son, to God showed the deity how deep and true was one's commitment. In Second Kings 3:27, the king of Moab took his eldest son "that should have reigned in his stead and offered him for a burnt offering upon the wall." He did this to win a battle. Sacrifice was serious to the people of the Bible. The more precious the offering, the greater the sincerity of the giver. In an age of possessions and descendants, the firstborn son, especially if he was an only son, was a supreme sacrifice. The firstborn son was the father's line to eternity, his mark upon the world, his hope, his salvation; and to give such a prize, such a treasure, to God was the epitome of devotion.

How much more important does it then become when a god sacrifices his son, his only begotten son, to man? How important does that make us when the gods sacrifice to us? This belief is at the very heart of Christian religion. It must be very satisfying to know that one is so important, so crucial to some great cosmic plan that a god would give up his most prized possession for our sake.

It is not the contention that Christ died for our sins that is important, but rather the fact that a god actually made a sacrifice to man. This was not just any sacrifice. It was not doves, not goats, or lambs, or bullocks—perfect in shape and form though they may be—but his son, his firstborn son, that of whom he is most proud, most loving, his greatest treasure. That's how important we are.

Animals were made for us. Birds fly in the sky for us. Plants grow for us. Winds blow for us. Rivers flow for us. We are important, we are crucial, and we are unique; and while we are very humble, we just must proclaim our uniqueness and our importance. We must reiterate our significance, basking in the rosy glow of satisfied humility.

It is enough. Our delusions are great, our illusions vast. We believe the improbable without doubt, the impossible without qualm. It is an astronomical misleading of the mind in pursuit of self-aggrandizement.

And yet reality presents us with powerful arguments to the contrary. Northern Ireland comes to mind. Here, two groups that worship the same god, accept the same savior, and supposedly try to live by the same set of rules cannot find anything better to do with their time than fight each other, killing women and children in the name of a merciful and caring god, all the while calling their war holy. India, the Middle East, Africa, and Southeast Asia have all spawned wars in the name of God. All kill and maim in the name of a benevolent deity. And no one seems to think this odd; no one says, "How can this be, the lowest of deeds for the highest of motives?" We all know, we all understand; it is the true nature of man. It is not pretty.

There is, however, one definitive statement that can be made regarding religion. For those whose piety is great, for those who believe in a hereafter, heaven and hell, and the durability of the soul, there is one irrefutable assertion that can be made regarding all religions that believe in eternity. If there is no such thing as life after death, no one will be disappointed.

Chapter 4

The only reasonable solution to delusion, especially religious delusion, is comprehension and understanding. If we know we are susceptible to myths, then and only then can we begin to address the problem. Barbara Sproul says in her book *Primal Myths: Creating the World* that when children are taught myths from an early age, they do not understand that what they are being taught is myth. They accept what they are taught as revealed truth. At the same time, these individuals reject what others have been taught as revealed truths. They are sure that any teachings except their own are myth. They are convinced strictly because of upbringing that they are right and all others are wrong. Thus are cultural myths preserved.

Religionists believe this way even when there is not one shred of objective evidence to favor their myth. They will accept their teachings as gospel with respect to their beliefs, but will reject the teachings of others with reference to other beliefs. In the final analysis, no one can justify their myth except by blind faith.

Blind faith that they are genetically predisposed to believe is revealed truth. Believers cannot understand that what they believe as truth others know to be myth because these others are sure they have revealed truth. Once all see that they have the same predisposition and that their own particular revealed truth is no different from all other revealed truths will we begin to see progress in rationality.

In the meantime, people believe they are right and everyone else is wrong. Never mind the lack of proof, the complete absurdity of such beliefs, or the predisposition to defend to death such beliefs. I am right, you are wrong. The song is old as is the predisposition, and it will not change until education outstrips belief, something no culture wants.

If we are to understand ourselves, we must begin at the beginning. We are, as a species, just one link in a long line of life that stretches back into the mists of time. We are evolved animals, and all our attributes, both physical and mental, have been impacted by the evolutionary process. To understand the human species, we must begin by understanding evolution. To appreciate what we are, we must first see where we came from. If we are to have any kind of a future, we must be aware of our past. We need to know why we are the way we are. To do so, we must first understand the evolution of life.

Why is there life at all? Why is there life in the universe? Life is certainly not necessary to the evolution of the universe. The stars, like our sun, could blink on, run through their allotted life span, and go dead or explode without any need to be noted. Planets could spin into being and circle in majestic isolation around said suns without any need to be populated. Vast seas and mighty mountain ranges could exist just as well without fish and trees and bears. So why do we have life at all? Unlike the elements, life is not really necessary.

It is possible that life may be an integral part of the universe. Recent experiments indicate the probability that wherever favorable conditions exist, life will occur. Science has shown that nature has the ability (without conscious motivation) to naturally form amino acids, urea, fatty acids, and nucleic acids from those chemicals that made up the early universe. On Earth, these chemicals, when heated, form proteinoid microspheres. Proteinoid microspheres can be called protocells because they can perform some of the basic functions of cells. Microspheres can grow by absorbing more proteinoid material. Microspheres can exhibit osmotic behavior, and they may even be able to multiply by budding or by dividing.

Are they life? It is too soon to tell, but in the next ten or fifteen years, we can expect to see one method whereby life from nonliving chemicals is pretty much elucidated.

Life on Earth probably began much as this very brief description outlines. Life on other planets may or may not have developed in the same way, but it appears that carbon-based life-forms can arise when common chemicals and potent energy sources, combined with millions and millions of years, all conspire together, unconsciously, of course.

We are talking here about two separate and different things. Evolution says nothing about the origin of the universe or the origin of life. Evolution is neither dependent on nor influenced by origins. Evolution can only occur after universes come into being and life originates. Evolution is an explanation of how simple life-forms become more complex.

Once life began, evolution took over. Evolution is really quite simple. The rules are twofold and straightforward: the first rule is survive; the second rule, reproduce. That is basically it. Organisms that survive and reproduce better than other organisms proliferate at the expense of those "other" organisms. Strategies that allow organisms to do a better job of surviving and multiplying tend to be perpetuated.

Pacific salmon start out as eggs in the bed of some wild-rushing stream. The eggs hatch, and the fry spend some time practicing survival skills in the stream. Otters intrude, kingfishers dive-bomb, and some salmon become food for other fish. Eventually the remaining fingerlings head downstream to the sea. Here they grow and fatten up and practice more survival skills. Seals love salmon, orca dine on salmon, and larger fish eat smaller fish.

After several years, a yearning for old puddles sets in, and now fully adult salmon swim back upstream to that very same stretch of water that nursed them as eggs and fry. Here the adults do the one thing they have been living and growing for all their lives— they spawn. The female lays the eggs—as many as six or eight thousand—and the male fertilizes them. The eggs settle into tiny cracks and crevices in the streambed to repeat the cycle. The adults die.

Along the Pacific coast of North America every year, millions of salmon swim up thousands of rivers to deposit billions and billions of eggs in cold, clear streambeds. Each egg is different; each egg will vary slightly, or not so, from all other eggs. And some of these variations will be helpful to the salmon egg that contains them. This gives these eggs just a little bit better chance of surviving and reproducing.

That is how evolution works. A species will produce far more descendants than the habitat can sustain. These offspring will compete with one another; less fit individuals will leave fewer or no offspring while fitter individuals will leave more offspring. And so the species is selected by nature in an ongoing and unconscious process to reach a point where each species is as close to the ideal for its niche as possible.

Evolution works by first creating genetic variation and then letting existing conditions determine who survives. The individual genetic variations available for selection are rarely perfect. Most times, selection is for something, anything that will assist the species in its fight for survival. The human eye is a good example of "it's not the best, but it does work," and so is selected for. The nerve fibers that make up the eye's optic nerve exit the rods and cones from the front of the retina rather than from the back. They then exit the eyeball through a hole in the retina that leads into the brain. As a result, all light entering the eye and falling on the retina must traverse a forest of nerve fibers to do so. This is a poor design, but after a fashion, it does work.

This is how evolution works. It is a make-do, catch-as-catch-can procedure. It even allows considerable leeway in selective response as long as such genetic predispositions do not endanger the species as a whole. Evolution via natural selection is not precise, exacting, limited, or even the best. The basic evolutionary theory is "if it works, use it." As long as the selection is even for a marginally more beneficial characteristic, it will be adopted.

The process also works in the opposite direction. A species can become very close to being perfectly adapted to a given habitat, provided that the habitat remains unchanged over a long-enough period of time. Now selection is for maintaining adaptations already made and selecting any deviations from the more or less acceptable adaptation. Evolution works to adapt and then to maintain a species adaptation to a specific niche.

Different species have different strategies for survival. Bears prey on salmon, eating forty or fifty fish per day in season in order to put on fat for the coming winter. But female bears do not produce thousands of offspring. Bear strategy differs from salmon. A female bear will give birth to one, two, or three cubs. She will nurse and care for these cubs for as long as three years. Even so, more than half the cubs will die—many at the hands of other bears, some in accidents, and some through starvation. In the long run, bear reproductive rates will just match the carrying capacity of their territory. Bear variation, like salmon variation, goes a long way toward determining bear evolution.

Bear and salmon, microbe and plant, it's all the same—survive and multiply. To survive, the organism must meet the criteria of natural selection. Nature selects based on ability; organisms that are better able to survive do so. Some insist that this is a tautology; only the fittest survive, but we define the fittest as those that do survive. Under that criteria, the statement is meaningless and cannot help us in developing a satisfactory evolutionary hypothesis. This would be true, except for two additional components of natural selection. The first component deals with variability. Each organism—as we have seen with the salmon eggs—is a bit different, and the difference sometimes offers a slight advantage in the fight to survive and reproduce. So natural selection is not random. Lives are not entirely won or lost on casting of the dice, but by a selection process that tends to preserve more capable organisms at the expense of less capable organisms.

It comes back to the central thesis for evolution. Organisms always produce more descendants than the habitat can handle. Some must die, many must die, some of predation, some of disease, some of starvation. Under terms of stasis, a given population of organisms will achieve optimum numbers for a specific habitat and will maintain those numbers.

The second component is the more important one. Organisms must not only survive, but also they must reproduce. Reproduction is the key because it is in reproduction that change takes place. And it is this change that eliminates the argument from tautology. "Survival of the fittest means that only the fittest survive" is a tautology only if no predictions can be made from the statement. Because the fitter individuals pass on their fittedness to their offspring, we can predict that any species, over time, will become more and more adapted to the particular niche that they occupy. We can also predict that the more adapted to a specific niche an organism is, the less likely it will be able to withstand a change in that niche.

We are the only species that has been able to multiply all out of proportion to habitat capacity. We have changed, altered, and reconstructed our habitat until we now number in excess of six billion. No other large animal even comes close. Rats, some say, may rival us in numbers. Certainly krill in quantities to support blue whales exceed our numbers, and ants also, as well as termites. But not much else and certainly nothing weighing more than a pound.

It wasn't always that way. At one time, we were as uncommon as any other predator. We lived and died on the plains of Africa for hundreds of thousands of years as nothing more than a scavenger/predator. And we were successful, too, spreading out of Africa into Europe and Asia. But we were no more successful than other predators. The lion and the leopard both colonized Europe and Asia, so that wasn't such a unique distinction.

Then things changed. We invented a way whereby we could survive and multiply in numbers beyond those imposed upon other predators. We began to experience surplus population.

Surplus populations occur when less fit individuals survive and reproduce. Under Darwin's theory of natural selection, only the fittest survive and reproduce. Man is the only species that has transgressed this law of nature. Surplus population is that portion of the population that exists in a specific habitat over and above the number of individuals that same habitat could support as hunter-gatherers. We will get into surplus population a bit later. Suffice to say, it is the key to man. It offers a possible explanation to any number of misconceptions, misunderstandings, and the problems involved in the nature and behavior of the human animal.

It all begins with evolution. We are like every other life-form alive today, the end product of a three-and-a-half-billion-year-old process. Like every other life-form alive today, we can trace our lineage back to the first faint stirrings of life on Earth. During that entire three-and-a-half-billion-year period, our ancestors' shape, size, and, most of all, behavior were molded by evolution; so it is with evolution that we must begin.

Evolution is the compromise that organisms have to make to survive. Life is a compromise, and like all compromises, it is less than ideal. An ideal organism would be able to adapt quickly and completely to any stimulus. An ideal organism would be able to defend itself from all danger, absorb energy from any and all sources, and reproduce under any circumstances. The compromise we see in evolution is the difference between the ideal organism and reality. There is no ideal organism.

Organisms do not mean to evolve. Evolution does not need to occur. There is no goal to evolution, no plan, and no purpose. Our belief that there is a goal to evolution and that human beings are that goal is a delusion.

This is not to say that organisms are not goal oriented. They are, but the goal is not to evolve. The goal is to survive and reproduce. This is what organisms do; this is what species do. Whatever it takes, whatever each individual is capable of, this will determine survival and reproductive success. And while it is not the conscious intent of the organism or the species, successful survival and reproduction are the basis for evolution.

Organisms do not mean to evolve, and some don't. Stromatolites are kinda neat. They look like oversize mushrooms, a cloud of domes sitting on black knobbly stems. They are thirty to forty inches high, twelve to eighteen inches across, and their tops are a sort of slimy gray green. Oh, and they are old—really, really old. Stromatolites are formed when cyanobacteria survive. So it is really the cyanobacteria that are old. In fact, cyanobacteria are, as organisms, 3.5 billion years old. The cyanobacteria of today are essentially the same as those that created stromatolites in the Warrawona group rocks in Western Australia 3.5 billion years ago. The same organisms were doing the same thing for 3.5 billion years. Evolution does not need to occur. Organisms do not need to evolve. A stromatolite starts out as a pebble or some other hard, fairly smooth surface on the seabed in the light zone. Cyanobacteria need light. They have tiny chloroplasts within their cells, and like modern-day green plants, these bacteria use chlorophyll to convert sunlight into the energy needed to grow and divide. With their chloroplasts and a smooth pebble, the cyanobacteria were able to dominate all the world's early tropical seas.

Cyanobacteria grow in mats. They divide and turn sunlight into energy, and divide and grow, and soon they have covered the pebble.

These bacteria also secrete a gel that helps protect them from the ultraviolet rays of the sun. The gel, however, is a mixed blessing. The gel does protect, but it also attracts very fine soil particles in the water. Even the clearest of water contains some of these particles, and they tend to get stuck in this gel. Over time, the particles build up and reduce the amount of sunlight reaching the bacteria.

Now the bacteria are not fixed. They do have some ability to move. When the gel gets cloudy enough to interfere with their ability to convert sunlight into energy, the bacteria literally swim up through the gel and commence operations in full sunlight. Meanwhile, the gel cements the bits of detritus to the top of the mushroom, causing it to grow by infinitely small amount. Over billions and billions of years, cyanobacteria have built up layers of stromatolites over 2,500 feet thick.

But growing and dividing and cementing do not come without a price. The atmosphere of the early Earth consisted of methane, carbon dioxide, hydrogen sulfide, and ammonia. By our standards, a pretty poisonous mixture. Then the cyanobacteria set up their little chlorophyll shops, using sunlight as an energy source and carbon dioxide as building blocks to grow and divide. In the process, the cyanobacteria gave off oxygen.

Oxygen at that time was a poison. Many contemporary organisms could not survive in the presence of free oxygen; their chemical processes were oxidized by the poisonous gas, and they died. Those that did manage to survive retreated to places where there was no free oxygen. But the cyanobacteria were not adversely affected by the gas—at least not at first—and continued to grow and divide and make mushrooms.

It is one of the great ironies of evolution that this simple, small, nondescript organism, growing and dividing over billions of years, building billions of mushrooms, neither wanting nor needing change, laid the groundwork for almost all other life on Earth. At the same time, it was putting itself in danger of extinction. The oxygen that the bacteria gave off as a waste by-product was the very same gas that made possible the evolution of more complex life-forms. Among the more complex life-forms to evolve were snails and sea urchins, which would evolve to, among other things, graze on the very bacterial mats that had made their evolution possible.

Cyanobacteria did not want change, indeed, did not change. But they were responsible for one of the greatest changes Earth has ever seen. These changes made you and me possible. As for the bacteria? Well, they still exist, still grow, still divide, still produce mushrooms, but only in places where they can proceed more or less undisturbed. One such place is Hamelin Pool in Shark Bay in Western Australia. The salinity of the pool is six times that of regular seawater and too salty for most marine life. Here, cyanobacteria continue to make mushrooms, totally unaware of what they have done, the changes they have made, content to grow and divide and survive—the same things they have done for the last 3.5 billion years.

Resistance to change is not limited to bacteria. On the plains of western North America lives another life-form—bigger, faster, and more complex but just as unchanging. For over a million years, pronghorn antelope have prowled the plains of the western United States.

Pronghorn antelope are superbly adapted to their arid environment. Primarily browsers, pronghorns get by on forbs, some grass-type plants, as well as cacti. They can survive with little or no water, getting all they need from the plants they eat. Their dun color and broken coat pattern are excellent camouflage for life among the fractured plains they call home. They have excellent eyesight and have fleet feet being able to hit top speeds of sixty miles per hour. Unlike the cheetah, the pronghorn has the lung capacity and overall physique to sustain such speeds for distances of up to nine miles.

The antelope's eyesight is also outstanding, believed to be the equivalent of eight-power binoculars. On the open plains, good eyesight is essential if an antelope is to see danger coming. Fleet feet are critical for avoiding same.

All in all, the pronghorn antelope is superbly adapted to its open-range home, having made most of the evolutionary modifications necessary to fit in perfectly. Sure, it would be nice to run a little bit faster, but as it is, pronghorn can outrun anything else on the prairie, and the cost must be considered. Faster means more energy, more food, more resources, which must come from somewhere. Speed is not free.

Pronghorn antelope do not want anything to change. Neither they nor their environment would be as well suited to each other if anything changed. As far as antelope are concerned, they do not want to evolve. And so the antelope all pray in their own antelopian way that nothing ever changes.

There is yet a third organism that demonstrates the predisposition of life to resist change. The environmental circumstances of this animal are considerably different from the antelope's and the cyanobacteria's. Indeed, one would think that these animals would fervently wish for better accommodations, would devoutly pray for less severe surroundings and slightly less strenuous living conditions. But they don't. They survive and reproduce in one of the most inhospitable places on Earth, a place we may find to be absolutely uninhabitable but that penguins find perfect.

It is May, and it is cold and black. Not the black of a normal night, but the black of a twenty-four-hour-a-day, seven-day-a-week night for months on end. It is the beginning of winter, and it is Antarctica. It is cold, as low as minus 140 Fahrenheit, and the wind regularly whips around the pole at speeds of more than 100 miles per hour. It is here on this cold black windy shore that emperor penguins have come to breed. Not only do they breed here, but also they have walked 120 miles through this black frozen waste to do so. There may be a more inhospitable, disagreeable ugly place and time for birds to breed, but it is doubtful, and the penguins will have it no other way.

The program starts in late March, which, down under, is the same as our northern hemisphere in October. The sun is going down and will not make any appreciable appearance for the next five months. The penguins have all waddled the 120 miles to the permanent ice shelf and their happy home for the next eight months. Immediately, the birds, which weigh sixty-five to eighty-five pounds and stand forty to fifty inches at the beak, begin to size each other up. Within five or six weeks, most have found mates, and breeding begins. Soon egg laying starts. Emperor penguins lay only one egg, and as soon as it is laid, the male takes it up on top of his feet and covers it with a fold of abdominal skin. He will remain like this without food—with the wind blowing, the snow flying, the surroundings pitch-black—for the next nine weeks. One consolation, though, emperor penguins nest in groups. This makes it easier to conserve body heat. Our male will try to get as close to the center of this group as he can. The result is a constant movement within the colony as birds on the edge try to get closer to the center of the flock and of warmth.

Meanwhile, the female has gone off to feed, waddling the 120 miles to the edge of the ice flow, a process that will take her two days. Here, she begins to feed and feed and feed. For the entire nine-week period she is gone, the male spends his time incubating the egg. During this time, not only does she put on weight, but also she is able to store semi-digested food in her stomach. Then, full to burst point, she trudges the 120 miles back to the colony, hopefully in time to see her chick hatch.

Mostly it works out all right, and at about the time she gets back, the chick is hatched. She, well stuffed after her trip to the ocean, takes over from the male, and he goes down to the sea to replace the third of his bulk lost during the preceding nine weeks. It is still pitch-black, cold, windy, and snowy, but things are getting better. Soon, like in three months or less, the sun will rise.

Our dutiful parents continue to trudge and feed their one chick as it turns from a small pile of fluffy down into a full-grown, if adolescent, bird. And at last, the reason for our emperor penguin saga becomes apparent. If the chick is to have the best chance of survival, it must fledge at that time when one of the world's worst places, weather-wise, is at its questionable best. That time is the first of December, when the temperature in Antarctica can reach a balmy 32 degrees Fahrenheit as a high on the warmest day of the year. This incidentally is distressing for the birds, built to take the very cold weather of the Antarctic winter. Thirty-two degrees is much too warm, so the whole mess trudges off to the ocean where the birds can stay cool.

It would be difficult, if not impossible, to deliberately design a more dangerous, cruel, and harrowing method of reproduction. And yet here it is, not only ongoing but also successful. There are over a quarter million breeding pairs of emperor penguins on Antarctica. They are so ideally conditioned, so superbly adapted to the cold and snow that they wouldn't have it any other way. Change would only do the penguins harm. If the Antarctic warmed up, the birds would be unable to tolerate the heat. If it got much colder than it is now, all life on the continent would be extinguished. So no change, please. Everything must stay just the same. Penguins do not want to evolve.

Organisms do not consciously or even unconsciously decide not to change. Antelope do not gather in quorums in the bowl of the plain to vote against change. They do not stand nose to nose to ruminate over the young chippy over there that has the nerve to be born with some cheap alteration. I seriously doubt that any given antelope on any given day could even spell the word *change*, let alone resist it.

No, what happens is this: Each antelope fawn is a bit different from all other antelope fawns. For 99.99 percent of the fawns, the "bit" is tiny and means little. Once in a great while, a fawn will be born with a little "bit" greater difference. Let us say that because of a slight change in muscularity, the animal will be able to run at sixty-three miles per hour rather than sixty, a 5 percent increase. The change comes at a cost, however. The antelope will have to increase its food intake by 7 percent in order to do this. In fact, it really has no choice. In order to survive, its food intake must exceed that of its contemporaries by 7 percent. It cannot say to itself, "I will eat what the others eat and go only sixty miles per hour." No, it has to consume 7 percent more in order to remain in good health, in order to procreate and pass on its genes for sixty-three miles per hour.

But it can only find enough food to increase its consumption by 4 percent no matter how hard it tries. As it grows up, it is always a few percentage points below what is required in order to remain in top shape and form. As a result, it is not going to pass on its genes. If it is a male, it is unable to compete with males in top form and is thus cut off from reproductive success. If it is female, it is incapable of giving its fawns sufficient milk, which will result in their being selected out. It is the mechanics of natural selection that has made our antelope tried-and-true. It is that same force that will keep them there.

*****

And yet evolution does occur. We know this from present-day observations and from the fossil record. There is within each species tremendous variation, a great deal of diversity. It is just that under the norm of natural selection, this variation is kept in check.

Dogs are a good example. All present-day dogs are descendant from either *Canis dingo* or *Canis lupus* or the common ancestor of these two species. From these two basic canine rootstocks, the dingo and the Asiatic wolf, over four hundred different breeds of dog have been developed in about ten to twelve thousand years. Most of these breeds have been created in the last four hundred years.

Big dogs weighing more than two hundred pounds, small dogs weighing less than five pounds. Long hair, short hair, no hair, curly hair, straight hair, red hair, black, golden, brown, white, and spotted hair—all from a basic gray rootstock. Dogs that track, dogs that herd, dogs that pull, dogs that hunt, dogs that guide, dogs that pro-tect—all these behaviors from one or two ancestors that were pur-suers of game over much of Europe and Asia about twelve thousand years ago. All this genetic diversity within the basic wolf stock. It is possible, if one so desired and had the time, to take an Asiatic wolf population and recreate all these breeds and many others besides.

And yet we don't see wolves with very long hair, short legs, multicolored coats, or different sizes. The wolf is continuously selected for just those attributes that will allow it to survive and reproduce at optimum levels. True, wolves vary in size and color, but that variation is limited. Not too big and certainly not too small, not too dark and not too light; like the baby bear's porridge, wolves are just right for their environment.

As we have seen, the reason we do not see extreme variation in organisms is that all organisms are adapted to a niche. Each has a place in nature that it has evolved to conform to. The emperor penguin is superbly adapted to the Antarctic shore ice. It survives; it procreates, and penguin populations are stable at the bottom of the world. Cyanobacteria in their stromatolite domes ruled the world until they pumped so much waste oxygen into the atmosphere, it made more complex life-forms possible. These life-forms eventually evolved to feed on cyanobacteria.

Each organism has a niche. The lion on the plains of Africa has a territory—defended, of course—from which to pluck passing zebra and wildebeest. The leopard, too, has its niche—different, seeking smaller prey mostly at night.

Those eaten also have their niches. Zebras are superbly adapted to the African plains and the tough browse that grows there. All these animals are adapted to specific niches. Plants too—trees, bushes, grasses, vines, all surviving and reproducing each in its own particular niche.

But this did not just happen. At one time, these animals and plants were not as well adapted to their niche as they are today. At one time, they were not adapted at all. But then things changed, and that is how all this adaptation to niches began.

It always begins with change. In the beginning, Earth was too hot for life, and then it changed. It cooled down; water began to liquefy and fill the lower levels of Earth. Life began, and soon the water teemed with simple cellular life capable of surviving and reproducing. Cyanobacteria were among these early life-forms. They survived and grew and domed much of the underwater Edens. In the process, they changed the world, becoming the first life-form to pollute the atmosphere. The history of Earth has been, since that first major change, just one change after another.

Our Earth is not static. Its condition is influenced by several factors. Climates change, forcing organisms to adapt or die. Plate tectonics pull landmasses apart, shove them together, and move them all over the globe in response to convection currents deep within the Earth. Climates and convection currents bring change.

But some of the most spectacular changes inflicted upon our planet come from outer space. In the last few years, it has become apparent that a huge meteorite or comet impacted Earth at a place called Chicxulub on the Yucatan Peninsula in what is now the Gulf of Mexico. It is the right age at sixty-five million years old. It is the right size at ten to twelve miles in diameter, and it left all the right clues. It killed the dinosaurs. That really changed Earth.

There is a clay layer that defines the Cretaceous-Tertiary boundary. This layer is filled with the signatures of a meteorite impact. There is a lot of iridium in the clay layer. Iridium is common in meteorites but uncommon on the surface of Earth. There is also soot. Soot from burning organic matter—matter that was set ablaze by the impact and heat from the meteorite. Then there is the shocked quartz. These grains of what was at one time sand show small fractures oriented in different directions. This kind of fractured quartz is typical of sand that has been subjected to a great deal of pressure and heat. Today, scientific opinion is that about sixty-five million years ago, a large meteorite struck Earth in Mexico, and as a result of that impact, all the dinosaurs died. A good thing, too, else we might not be here.

It wasn't just the big dinosaurs—it was all the dinosaurs and all the flying reptiles and all the marine reptiles as well as all the ammonites, a truly awesome extinction, and niches opened up all over the place.

Mammals, who had been around as long as the dinosaurs but were kept to the sidelines by these overpowering reptiles, now came into their own. They filled the empty niches. They radiated, they diversified, and they occupied all the openings left empty by the dinosaurs. But they could only do this because the meteorite just happened to be at the right (or wrong for the dinosaurs) place at the right time to intersect Earth. And it was all a matter of chance.

Students of human nature do not like the idea of chance. Intelligence is inevitable, they aver. Evolution will eventually lead to intelligence. Well, maybe yes, maybe no; it is still a matter of chance. We are not as a species predestined. Our evolution, as we shall see, was a result of random acts that could have just as easily not occurred or have occurred with different results. Our inevitability is one of our more cherished delusions. It is constantly presented with the proposition that if there is life on other planets, then there is intelligent life on those selfsame planets. Nothing could be further from the truth. Intelligence is so unusual, so unlikely that it is only under very rare, very unique circumstances that it will ever occur.

Claws are the answer. Pointed, powerful, and, most of all, large—these survive. Speed, speed, and more speed—this survives. Fangs—long, sharp, and potent—these survive. Camouflage, stripes, patterns, colors—these survive. But intelligence, not a chance (there's that word again). Only under certain unusual, unparalleled circumstances does intelligence play a part in survival. And then it is an extremely risky adaptation unlikely to succeed. So while there may be millions, if not billions, of planets in the universe populated by various forms of life, there will probably be less than one intelligent life-form per galaxy.

But still, we are here, and we claim intelligence. And so the obvious must be dealt with. Like all life, we are the result of chance and evolution. It always comes back to the same thing; we are the product of our past. What happened in our history allowed us to survive, unlike more than 95 percent of life on Earth. We survived by luck and by behaving in ways that improved our survival and procreative potential. Just as all animals behave, we behaved in ways that helped us survive. Nature tells us that certain behaviors enhance survival as a species, and we listened.

How did we survive? No fangs, no claws, bare-assed naked, puny with only the poorest of assets. The story is a fascinating one—an epic of chance, opportunity, and, as we shall see, two advantages. First of all, we had a place of refuge in which to evolve, and secondly, we developed the ability to control a chemical reaction. It all had to do with fire and water and improbability and a huge dollop of chance. We did evolve, but we could have just as easily gone extinct.

*****

We have seen that evolution does not need to occur. Some organisms are essentially the same as their ancestors of three-plus billion years ago. We also know that organisms do not want to evolve. We know that there is no conscious or even unconscious predisposition on the part of individual life-forms to change. If organisms could control their destiny, they would vote en masse for stasis. Let nothing change. Let everything remain the same, thank you.

And yet evolution does occur. The fossil record constitutes overwhelming evidence of evolution. We have found in the last twenty years so many instances of the evolutionary process in the bone beds of the world to demonstrate conclusively that evolution has occurred.

In China, paleontologists have found dinosaurs with feathers. In Europe, paleontologists have found birds with teeth. The inescapable conclusion is that birds evolved from small bipedal tree-dwelling dinosaurs, and the fossil record demonstrates this beautifully.

The feathered dinosaurs from China could not fly. The original purpose of feathers, which are nothing more than modified scales, was to keep these small chicken-size animals warm. It was only later when some of these feathered dinosaurs became arboreal, racing about the treetops in search of insects, that the additional potential became apparent. Dinosaurs that could move from tree limb to tree limb by extending their feather-covered forelimbs and gliding improved their survival chances. And so birds were born.

The evolutionary history of whales has also been documented. From coyote-like ancestors, our modern whale went through a series of intermediate steps to reach their present size and shape.

Whales are believed to descend from a group of mammals called mesonychids. The mesonychid line, now extinct, consisted of animals that were hoofed like cows or hippos but had teeth meant for meat eating. Early whales had teeth that looked much like those of the mesonychid's.

Among the earliest of these that we know of is *Pakicetus*. *Pakicetus* looked like a cross between a coyote and an otter with a bit of sea lion thrown in. There is evidence that *Pakicetus* was active in the water as well as on land. It had four legs, a tail, a large skull with sharp teeth, and very odd ear bones. These bones consisted of two shells that looked like small grapes. These shell bones were connected to the animal's skull by bones in the shape of an S. Only whales have such ear bones. No other vertebrate has them. *Pakicetus* was not in any way, shape, or form a whale, except for those darned ear bones.

Next we have *Ambulocetus* or walking whale. Still not a whale, it was long and squat much like a mammalian alligator. Like modern alligators, it spent much of its time in the water but came out to sleep, to mate, and to give birth to its young. *Ambulocetus* was not by any stretch a whale, but it was on the way. A heavy body and big tail coupled with powerful hind limbs made it a superb swimmer. And then there are those whalelike ear bones.

Next comes a series of fossils. *Dalanistes* had heavier hind legs, a squatter frame, and cetacean ear bones. A bit later, we have *Rodhocetus*. Adapted to a life in the water, it had hind legs, but they were barely connected to the spine, and it is almost a certainty that they would have been of little or no value on land.

Then come *Takracetus*, *Gaviocetus*, and *Dorudon*. All are stream-lined, all are whale shaped and with hind legs—useless hind legs. Hind legs very poorly attached or not at all attached to the vertebra.

One could not ask for a better intermediate between whales and their past than *Basilosaurus*. *Basilosaurus* grew to lengths of fifty feet. They had a slim rangy body, tail flukes, a small head, sharp teeth, and tiny hind limbs—hind limbs only a few inches long but complete right down to five dainty, diminutive finger bones. *Basilosaurus* is the last whale with legs.

The picture is not complete. The arrangement of these animals in the geologic column is not absolutely certain. Some of them may not even have been ancestors but were rather cousins that dead-ended. There is, however, overpowering evidence that whales did evolve from a set of ancestors that lived and died on dry land. The ear bones tell us that.

The list goes on and on. We have a pretty good idea of how horses evolved. We have a good handle on the evolution of mammals from early mammal-like reptiles. Trilobites offer excellent evidence of evolution, showing time and time again how one group evolved into another.

We also have an outstanding, if somewhat limited, list of human ancestors—beginning with *Ardipithecus ramidus* at almost five million years old up through *Australopithecus anamensis*, *Australopithecus afarensis*, *Australopithecus africanus* to *Homo habilis* and *Homo ergaster* to archaic humans. And in the last two hundred thousand years, we see the emergence of modern man. We will deal with our own evolution in more detail a bit later, but suffice to say, we have a pretty good idea of human development up to a point.

A split from chimp-like ancestors took place six to eight million years ago. We have found no fossil hominid remains from this time for reasons that will become clear when we pursue the ancestry of man.

So evolution occurs. We know that the fossil record tells us that. DNA tells us that. And the relationships among today's animals tells us that. This time the question is not so much about why but how, and in the how we can begin to see the why.

The history of Earth—with its ebb and flow, its heat and cold, and its pushing together and its pulling apart—has had a huge impact on evolution. And as we have seen, so do impacts from outer space. However, for much of life's history, the condition of nature has been stasis.

Stasis is a key to understanding evolution. Over a billion or even a hundred million years, Earth will undergo substantial change. Continents will drift, volcanoes will spew change throughout the land, meteorites will impact our orb, doing greater or lesser damage to earthly life. But in terms of millions or even tens of millions of years, there will be relatively long periods of stasis. Dinosaurs came into their own during the Triassic and became the dominant animal group for the next 185 million years. Now no single dinosaur species survived over that period. But in most places around the globe, conditions remained stable enough that these reptilelike creatures could continue to exist—a period of stasis when evolution did occur but only within limits.

The beginning of the era of the dinosaurs was characterized by an event of some kind that caused the extinction of 90 percent of the life-forms on Earth. This was the famous Permian-Triassic extinction. The deaths of so many plants and animals opened the door for the dinosaurs. Then sixty-five million years ago, another mass extinction event closed that very same door, and all the dinosaurs died.

Evolution takes place under two conditions. It occurs during periods of change, and it occurs during periods of stasis. During stasis, any new organism that has to compete with existing organisms for survival stands a very good chance of losing. Most organisms that survive quickly adapt to their habitat, the pronghorn antelope being a good example. Because of the difficulty new species have competing with existing species, new species rarely appear during times of stasis. As a result, there may be periods of tens of millions of years where little or no evidence of evolution shows in the fossil record. Species that are more or less adapted to their niche may, over a period of time, give rise to genetic mutations that are slightly more in tune with their habitat. But most adaptations are made quickly, the selective process being biased against any change that does not contribute to adaptation.

We may see slight modifications of organisms over long periods of time in response to changes in conditions. Continental drift is a good example. As the landmasses move over the surface of Earth, their location at any one drift point will influence the climate of that landmass. Drift southward or northward from the equator, and the climate will change. Drift far enough, and it can get downright cold. The selective process then becomes one for thicker fur, fat layers, and shorter extremities to reduce heat loss. If the landmass goes the other way, then just the opposite may occur—longer, skinnier bodies, thin fur, and no fat.

This is the type of evolution that occurs under relative stasis. It is slow, limited, and not prone to leave many clues as to its occurrence in the fossil record. So evolution occurs during periods of stasis such as we are now experiencing, but it is hard, if not impossible, to observe over the period of a human lifetime.

Evolution during periods of change is different but would still be hard to observe during a single human lifetime. Adaptation is quick in evolutionary terms, but that still means tens of thousands if not hundreds of thousands of years. From an evolutionary standpoint, under conditions of change, evolution is rapid and pervasive. At the end of the Permian-Triassic die-off, niches opened up all over the place. Now a different selective process began. An organism had only to have a minimal adaptation for a given habitat. Almost all the habitats were empty as a result of the die-off, and it was not necessary for organisms to try to establish themselves in competition with already existing organisms. Even so, adaptation to a specific niche took thousands of years.

As we have seen, evolution occurs during stasis, but it is slow and rare. We have evolution during change where it is rapid and robust and shows up very well in the fossil record. But there is a third kind of evolution that goes on all the time, and it is a very different kind of evolution. It is called loser evolution, and it does not work very often or very well, but when it does, it produces evolutionary exceptions.

Loser evolution can occur in any given habitat. Habitats are not uniform. Some places within the habitat will be better than other places. The organisms that exist in these superior places have an advantage over the same kind of organism in a less desirable place. It will be easier for these organisms to survive and reproduce. Perhaps there is more food, perhaps a better water source or more hiding places, anything that would allow a marginally better potential for survival and reproduction.

There is a feedback mechanism involved here. Organisms that inhabit more desirable habitat areas are a bit stronger and a bit fitter than the same kind of organisms that occupy less desirable parts of the habitat. The organisms that occupy the more favorable parts are in better shape and are thus able to defend and maintain their desirable patch of ground.

Less able organisms are regulated to the less desirable parts of the habitat. They live in more marginal circumstances and are not quite as fit as their more fortunate brethren. As a result, these organisms are unable to fight their way into the more desirable habitats. They are losers.

Most will live a more or less precarious existence, eking out a living, surviving, and reproducing but not as well as their more fortunate brethren. They are always more exposed to any adverse circumstances that might arise. Should drought come, these organisms will die sooner and in greater numbers than their more fortunate neighbors. They are at greater risk of disease because they are not quite as healthy, at greater risk of predation because they are not quite as strong, and at greater risk of lower reproductive rates because of less-than-ideal surroundings. They are also those organisms that have very slim but very real potential to dramatically change the dynamics of the evolutionary process.

Humans, indeed, all terrestrial animals and plants originated in the sea. In some shallow sun-warmed pond, it all began. For many millions of years, that's where life stayed, plants and animals surviving and reproducing and evolving in the seas of the world. Over a period of time, the seas became crowded, the bays full, the estuaries overflowing, and the lagoons loaded with life. But on land, it was a different matter. Plants and insects abounded, but no vertebrate animals were to be found. Land was a pristine paradise, and the sea was crowded. Even so, colonization of the land by fish was slow, reluctant, and haphazard. Animals came ashore not because they wanted to but because the alternative was extinction, loser evolution in action.

Picture a pond—wide and shallow, surrounded by trees, and not very deep, say, eight to ten feet in the middle—gradually shading up to the shore. Trees are the key. They are large, tall, straight trunked, and have a bad habit of dying and falling into the water. As a result, the rim of the pond is a hodgepodge of tree trunks, limbs, and branches—a virtual wood wall extending some one hundred feet or so out into the pond. It is an absolute labyrinth of wood—dense, almost impenetrable by fish, and poorly stocked with easily edible prey.

The pond is occupied by a species of lobe-finned fish or maybe even two or three species of lobe-finned fish. Lobe-finned fish have fleshy lobes to which their pectoral fins are attached. In some species, the fleshy lobes contained bones among which scientists have identified the humerus, the ulna, and the radius—all bones of modern vertebrates but not normally found in fish, a fish with arms or at least the start of arms and selected for too.

The fish in the center of the pond live the good life, relatively speaking. Food is easier to procure, the open water makes danger easier to see, and the center of the pond is less likely to dry out in the occasional droughts. All around, it is a better place, and the dominant fish live there.

In the shallows, less combative fish reside. But life in the shallows is tough, and a lot of it has to do with the trees. They are big and long, and since the ground on the pond side of the shoreline is wetter and weaker than the firmer ground farther inland, trees tend to fall out into the pond. Here they form long obstructions to movement around the edge of the pond. A four-foot-thick tree trunk in three feet of water forms a barrier no fish can overcome, unless that fish has a set of armlike pectoral fins—small and inefficient fins but perhaps just good enough for them to be able to scramble over a log to escape a predator or find new richer feeding grounds. And a fish takes the first step toward life on land.

Did it happen that way? Probably not. Another theory hypothesizes that periodically the ponds dried up and everyone had to go, or go. That would shoot the loser theory all to hell. So I don't like it. But it may be true, and it was only loser lobe fins that lived in pools that dried up.

As we shall see, we are the product of loser evolution. We lost out originally to our chimp-like ancestors and were forced back into the sea. There we multiplied, and soon, it was the seaborne losers that were forced back onto the land. These losers over time became winners, became us, and, in the process, destroyed our seaborne brethren. We no longer live in the sea. But old habits die hard; everyone likes to live next to the water.

CHAPTER 5

There are two primary characteristics that define social animals. The first and most important is dominance; the second is the territorial imperative. Territory is crucial to most social animals. Acquisition and defense of territory is a central part of social life. The goals of life are simple—survive and reproduce. To survive, an organism must have access to food and shelter. For some, the only way to secure sufficient food and some sort of reasonably secure resting spot is to acquire and defend a territory containing these essentials to survival.

Lions are territorial animals. The makeup of the pride is centered around the females and the cubs. Males come and go, defend the pride's territory, and sire offspring. The pride occupies a territory, a clearly delineated and defended space large enough to provide the group with sustenance. Part of the male's job is to patrol the territorial boundary and mark it with a particularly pungent spray of urine, the same pungent spray domestic male cats use to mark their perceived territory.

Lions require a territory, an exclusive piece of property needed to provide food for the pride. Too many lions would decimate prey species, and all the lions would suffer. By acquiring and defending territory, predators ensure that they have a reliable food supply. This may mean that lions with no territory do not survive. But the death of some lions is better than poor health for all lions.

Chimps are territorial animals. They will acquire and defend a patch of ground to the extent that males will seek out chimps in adjacent troops and kill them. Hyenas are territorial, as are leopards and baboons. We have territorial fish, territorial birds, and even territorial reptiles. The genetic predisposition to acquire and defend territory is common in the animal world. But an even more important characteristic of animal life is dominance.

Dominance is a recurring theme in nature, and the reason is quite simple. That is to make optimum use of reproductive potential all females of a given species need to participate in the process. This ensures the maximum number of progeny per breeding season.

Male participation is another matter. Under ideal conditions, a single male of a given species would be best suited to fulfill all the duties of a progenitor. One male has all the attributes necessary to sire offspring with the best survival potential. Okay, maybe two or three or even four males are the best, but if the parameters are fine-tuned enough among the three or even four, one will be marginally better. So under ideal conditions, it is one male and all the females for the best selective results, and the hell with the rest of the males.

It doesn't work that way. It would be impossible for a single male elephant to service all the hot-to-trot female elephants in Africa. Not only would the beast be worn to a frazzle, but his equipment would also be, let us say, overextended. This same situation would be true for most organisms—buffalo in North America, llamas in South America, lions in Africa, or even kangaroos in Australia. All these species would benefit if a single male sired all the offspring. But in all these cases as well as in all other instances, this is not possible.

But organisms that aspire to this ideal have better survival potential. In those cases, males contend with one another to determine dominance. Rank is established, and the dominant males service the females to which they have exclusive entrée. This is as close as natural selection can come to the ideal of one male and all the females. This is why dominance occurs in most social species.

Dominance predisposition is deeply woven into the genetic fabric of all animals. Reptiles, fish, mammals, birds, and even insects all demonstrate dominance predisposition. Anthias, *Serranocirrhitus latus*, is a species of small fish that carries dominance to its logical extreme. All anthias in a school are born female. But male input is required to fertilize the eggs, and the story of the missing male is one of nature's best.

Anthias schools consist of fish of various ages, and the established dominance system is based exclusively on age and, hence, size. The whole school is dominated by one male. But where did the male come from if all the anthias were born female? Well, at one time, he was a she, contributing like all other shes to the production of eggs. When the dominant male dies, the next anthias in line, a female, undergoes a physical change from female to male. Sexual organs that are female whither, and in their place, male sexual organs develop. So after several years of contributing only a number of eggs to the school gene pool, she becomes a he with the responsibility of fertilizing all the eggs. And thus passing on her or his dominance predisposition to succeeding generations. All this is accomplished through dominance in anthias society.

Closer to home among primates, we also see dominance systems in place. Gorillas have gone the way of the harem, one silverback male with several female companions. As male gorillas reach adulthood within the silverback's group, he drives them out. They, in turn, lead a bachelor life until they mature to become silverbacks in their own right at which time they will challenge the harem master for leadership of the troop.

Among chimps, a somewhat different hierarchy presents itself. Troops consist of mature as well as immature males and females. Both males and females establish dominance hierarchies. Males, many times, form alliances with other males, each supporting the other when challenged. Single males have little or no chance of overcoming such a coalition unless they, too, form mutual support relationships. Dominance is common in chimps; even bonobos, the sexiest of chimps, establish a dominance system among both male and female lines. They do this because dominance has survival value.

More dominant members of a species have better prospects in everything. Dominant animals take food from less dominant animals. The best sleeping places, the best resting places—all go to the more dominant individuals. As a result, their lives are marginally easier. But the main benefit of being the dominant male animal in chimp society is that females are attracted to dominant males.

When a female chimp comes into heat, she will be serviced by all the males in turn. But at that time when she is most likely to conceive, she sometimes runs away with one of the more dominant males, and the two will copulate numerous times to the exclusion of other males. As a result, because dominance behavior presents males with increased reproductive potential, it is maintained within the chimp troop.

How common and extensive is dominance among male chimps? One set of observations done recently by primatologists leads to the conclusion that one-third of all chimp males are killed by other chimp males. Dominance is an important part of evolution. It is an important behavior among chimps, both male and female.

Dominance also occurs in insects. In Southeast Asia lives a bug, but not just any bug. At five inches in length and three and a half ounces, this is one big bug. It is called the Atlas beetle (*Chalcosoma atlas*) and for very good reason. The head of this beetle is adorned with three large horns, two on the upper part that are fixed and one on the lower part that is moveable. Between them, these horns act like pincers and are used by the males for jousting with other males. Dominance as a method for determining survival and procreative potential is alive and well among the world's insects.

Atlas beetles live in rotting vegetation and logs in the jungles of Malaysia and Indonesia. Female Atlas beetles lay their eggs in rotting logs, and the larvae feast on these fallen banquets. Males vie and joust with other males to occupy and defend the most desirable spots on the rotting logs. More dominant males occupy the best food sources and attract the most females. The result is beetle offspring, with the male's predisposition to take and defend desirable food niches. Dominance survives because dominance works even among insects.

It is a lazy Sunday afternoon—church is over, biscuits are in the oven, chicken in the pan, potatoes on the stove, and all is well. A blue sky laced with fluffy clouds canopies the house. Flowers bloom; the garden, in its stately rows, exhibits full production. The sun shines brightly around the white clouds, and a soft cool breeze comforts all.

The farm is quiet in the afternoon warmth, and small boys explore after accepting, reluctantly, admonitions not to get dirty "because those are your only Sunday clothes." Small boys rarely listen.

A fascinating enclosure worthy of inspection is the henhouse. This structure consists of a wooden shed, at one end of which is an enclosed wire pen made from, of all things, chicken wire. The shed is where the hens sleep on roosts. It also contains small enclosed boxes where these same hens deposit their eggs. The hens want to produce chicks, and the farmer wants their eggs before they turn into chicks. So the chickens lay, ever hoping to acquire enough eggs to brood, and the farmer allows the hens one rooster to further fool the hens into thinking that someday, some way, they might achieve their heart's desire.

The rooster is the first point of temptation for young boys. Largish, red feathered, and with an impressive array of wattles and a bright red comb, the rooster sports an attitude. The hens will waddle away, clucking at the sight of little boys, but not the rooster. He is up against the chicken wire in a flurry, wings spread, feathers ruffled, and a sharp beak ready to peck. The rooster will then run up and down the inside of the pen as if searching for some way to get at these meanies who delight in tormenting the cock of the rock with twigs and short sticks.

The rooster is interesting, but the hens are the real attraction. Whispered discussions are held, furtive glances are directed at this hen or that, reasons are discussed, passionate arguments made sotto voce about the fate of a specific hen. Chicken was served every Sunday at the farm, chicken that came from the chicken coop, and curious boys wonder which will be next. Which one? There are clues.

The hens in the pen are all the same—red feathers, smaller wattles, and none the size of the cock. But they are also different. Some are fat, sleek, and very self-confident in a chicken sort of way. Others are thin and act somewhat harried. One bird, among the smallest, is the most neurotic hen in the coop. Even her feathers look bad, dull, matted, and, in spots, gone, leaving behind an embarrassing amount of bumpy pink skin. She is at the bottom of the pecking order. Will she be next Sunday's meal? Little boys want to know.

"Bottom of the pecking order" has come to mean in English exactly what it means in chicken behavior. This individual is the one that everybody picks on and has no one to pick on. Chickens insist on establishing dominance hierarchies. It is innate, it is hardwired, and it is what chickens do. Alpha chicken pecks every other chicken but does not get pecked by any chicken. Beta chicken pecks all other chickens except the alpha. Gamma pecks all other chickens except the alpha and the beta and so on. Each chicken in the flock knows who is "peckable" and who is "impeccable." And little boys learn, over time, that it is always the little hen—pecked and bedraggled— that gets the axe.

Dominance is not limited to the hens. There is a reason that the flock only contains one rooster. Two roosters in the hen pen would result in the death of one of them. In Asia, Central and South America, and, until recently, the United States, cockfights were considered to be an excellent form of entertainment. Two birds, their natural spurs further enhanced by longer, sharper steel ones, are placed face-to-face in a small pit. One will kill the other. Spectators consider this to be fine sport.

Dominance is common among birds, not just chickens but among many flocking birds. It has value, it sorts and separates, and it keeps the species healthy and better able to cope with survival and reproductive problems. Stronger, tougher birds make better parents, produce healthier chicks, which share, among other things, the ability to pass on the predisposition toward dominance.

On the plains of Africa, there lives the largest land animal in the world. African elephant bulls can measure over twelve feet at the shoulder and weigh more than ten tons. Both males and females establish dominance hierarchies, although the females' ranks are looser than those of the males'.

Elephant social life is built around the cows and calves. A herd consists of females, calves, and adolescent offspring. When a bull calf reaches the first stages of maturity, he is driven from the herd to live a life of near solitude. A bull is allowed to approach the females only when he is fully adult and has demonstrated a dominant position in the male hierarchy. Female elephants prefer to mate with dominant bulls because offspring of such bulls have a better chance of becoming dominant and of passing on their genes. Such bull potential also has benefit from a fitness standpoint. These animals are stronger, healthier, and larger. All this improves the survival and procreative potential of a calf. And so a predisposition to dominance is maintained within the elephant society for the very simple reason that it has survival value.

Beneath the hard packed sands of the African desert lives another species of dominance-oriented animal, the naked mole rat. The name is descriptive; the animals are hairless, they live like moles, and they are indeed rodents. Naked mole rats are three to four inches long and live their entire lives underground. The social unit is divided into three groups—a queen, her consorts, and the diggers. All the offspring of the group are born by the one female queen. It is her total dominance of the pack that allows her to do this. The queen has as many as three male consorts with whom she mates. She also has a small number of largish males who act as soldiers protecting the pack. The queen asserts her dominance by stressing all the other females in the group. She stresses them to a point where they cannot ovulate. And thus maintains her monopoly on offspring.

Dominance is common in nature, and it exists because it has survival value. One point: the predisposition toward dominance must exist before it can be used as a survival mechanism. At one time, some organism began to exhibit dominant behavior. In that organism's genetic makeup was a code to act in such a way to dominate other organisms of the same species. This may have occurred several times before the female genetic response clicked. Sexual attraction can perpetuate a predisposition toward dominance, but it cannot originate such a predisposition. The need to dominate is much deeper than sex, for it is only after it is manifested that it can be selected for by sex.

The sound is weird, unearthly, mechanical, definitely not of animal origin. It is a combination of a really, really big fingernail scraped across a blackboard coupled with someone letting air escape from the constricted neck of a very large inflated balloon. All this is followed by a series of loud grunts. It is as alien a sound as you are likely to hear, and every year thousands of people flock to Rocky Mountain National Park to do just that.

It is the sound of a bull elk in full dominance display. Elk mate in the fall; calves are born in the spring, which gives the young animals all summer in which to grow and put on fat for the coming very lean winter. Bull elk are harem masters, with one male monopolizing the lives of anywhere from two or three cows to as many as twenty-five or thirty cows. Bull monopolizing is severely restricted, however, and only the strongest, healthiest, most powerful bulls will acquire harems. And so bulls challenge and fight and determine dominance. Before the bulls can fight, they must first meet, hence, the bugling. The signal is well understood in elkdom: "Come and see how strong I am, how big my rack is, how powerful I am. I have many cows. To get to them you will have to come through me." So be it. Dominance, writ large, is pervasive among elk.

Hyenas offer us a somewhat different dominance order. Like the naked mole rat, the African spotted hyena has a social structure that is female dominated. Females are larger than males, stronger than males, and are the leaders of the hyena pack. Contrary to common perception, spotted hyenas are not cowardly scavengers. They actively hunt and take down over 80 percent of their prey. On the other hand, they will eat anything from insects up to and including other hyenas.

Females not only lead the pack, but also they lead the hunt, the choice of dens, and the defense of territory. They are the dominant individuals within the group, and this dominance has manifested itself in an unusual way. Female hyena genitalia bares a strong resemblance to male hyena genitalia. The clitoris is much extended and resembles the male penis; there is even a fat-filled extension that resembles the male's scrotum. This is not something that could be selected for in hyena society. The females pick and choose, not the other way around, so male preference would have no bearing on the matter.

These particular anatomical characteristics are an effect, not a cause. They exist for the same reason female hyenas dominate hyena life. Testosterone, a hormone that in males fuels dominance, also produces certain physical changes. It is high testosterone levels in female hyenas that produce both dominance and physical stimulation to certain parts of the female's anatomy resulting in male-like sexual appendages.

Dominance hierarchies serve yet another function in the lives of social animals. Dominance provides an order to the social structure of the group. Dominant individuals make decisions about which direction the troop would travel, when to go for water, and which sleeping platform is to be used. Less dominant animals acquiesce, and order is maintained. The troop enjoys a degree of tranquility because every decision is not fought over. So a dominance system provides a social species with selective advantage. Animals that behave in ways that promote dominance have an advantage over species that do not. Over time, it is these dominance-oriented organisms that survive and procreate.

We are social animals. We have dominance systems built into our social fabric. Every culture that has ever existed is a dominance system. Every culture that has ever been described says, "If you do the following things, you will be looked up to, admired, respected, and given status. If, however, you do not follow our laws and rules and customs and manners, we will look down upon you, criticize you, judge you unfavorably, and possibly kill you." Within the social group, this is the common thread that ties everyone together. Individuals know who is the leader and who are the followers. Things are done; actions are taken in accordance with rank. Power and responsibility goes with status, and esteem is accorded to high-ranking individuals. Each member, especially the males, devote a great deal of time and energy to improving status. At one time, this was accomplished mainly through strength. But now it is accomplished as much through guile and intelligence as it is through strength. And increasingly, the role becomes more and more complicated, more and more difficult to achieve. And yet we aspire, we try, we scheme and plan to achieve dominance within our social group. Always it is dominance.

Dominance drives us; delusion blinds us. Both of these factors are predispositions, and together they explain the very basis of human behavior. Dominance predisposition is not universal in the human animal and is generally confined to a selected few. It is the behavior of these selected few, however, that impact all of us.

Delusion is universal in the sense that we all believe without a shred of evidence to support that belief. Not only that, but also we consider it a right and duty to inflict others with our beliefs. Dominance has almost single-handedly brought us to our present state. Delusion has been the great brake on human potential.

While not as common as delusion, the predisposition toward dominance does occur in the human animal; we see it every day. In sports, there is a ranking, a hierarchy, winners and losers. Sports teams are small and for good reason. Nine men in baseball (there are more, but it is the playing nine that is the core of the club, and it is this group that will have a disproportionate impact on the ranking of the team). Football has eleven men or twenty-two, counting offense and defense. Soccer, rugby, basketball—all teams, all competing, all out to dominate the opposing team.

And we all know this; we understand and we encourage. From the soccer mom screaming at her kid on the local field to the Super Bowl, a venue measured lately by the cost of a thirty-second spot. It's all dominance. The World Series, the Olympics, the playoffs, contests all pitting team against team, person against person, to see who will win, who will triumph, who will dominate. Dominance is the rule in sports, but it is also found in other areas of human behavior.

In business, two companies making the same product will compete and try to dominate each other. They will try to make a better product than their rival. They do this to make more money. They want to make more money because we all agree that the accumulation of wealth is a desirable goal, and people with lots of assets are more dominant than individuals with fewer assets. This is true in all cultures even those with "different" assets.

Then there is politics. Here is dominance with a brazen face. Who controls who, who outranks who, and who gets an assigned parking space at the national airport? Consider also the dictators; they are dominance writ large and for no other purpose than to dominate. We will get into more detail on this later, but suffice to say, dominance drives us—some of us but not all of us—and the question is why?

The predisposition to dominate knows no limit. Time and again, we have seen individuals try to conquer the world, to rule over all, to dominate everyone and everything. Individuals with this predisposition spend tremendous amounts of energy attempting to reach their goal. But while the predisposition may be boundless, it has only been recently that dominants have had the ability to project themselves beyond their own clan.

Until approximately ten thousand years ago, we lived in small groups of perhaps 30 to 40 individuals. We defended a territory that could be as small as 20 miles by twenty miles up to an area as much as 150 miles on a side. The amount of territory defended depended on the relative affluence of the land. The operative criteria was sufficient food and water to support the group.

Males (and females) established dominance hierarchies within the group. This was an ongoing activity. Males were constantly measuring each other to determine rank. Bonds and relationships were established, broken, and then rebuilt as new individuals came into the ranks from the children of the females and as males grew old and infirm. This posturing and challenging was not a constant thing. The alpha was top dog; but the beta was very close behind, and the gamma at third is only a year or two from challenging the alpha. The result was a balance—the dominance structure built on a limited truce. Each individual knew his position and watched for signs that the individual next higher on the dominance ladder might show some sign of weakness.

At the same time, cooperation among all adult males was needed to maintain troop strength and to hunt big game. So each male knew intimately every other male. They knew each other's strong points, weak points, capabilities, limitations, emotional state, and physical capacity. Who did what in the hunt depended on this knowing. Who was good at stalking, who could set up an ambush, who was the most knowledgeable about game. Who would stand and fight, who would run (running away wasn't good, and even timid animals were tempted by a retreating back into thinking they were omnipotent). So these dominance hierarchies were balanced enough, contesting to determine dominance but not so much that the group was torn apart.

So dominance was an ongoing thing, a constant contest. Subtle moves and actions were designed to influence opinion. Mostly visual, a bit physical, it had to be done in the presence of the dominated, and so dominance was limited to the social group. Dominance could not be projected beyond the group. There was no way an alpha could dominate two groups, even if he was a super alpha.

But the need to dominate could be very strong. The predisposition could be in and of itself much stronger than necessary because no matter how strong it was, it would be confined to the group.

Dominance on any scale greater than the group could not manifest itself until about eight thousand to ten thousand years ago. With the advent of farming-herding, a number of changes allowed dominant individuals to extend their influence over greater and greater numbers of people.

Hunter-gatherer groups have relatively simple cultural interactions. The two main determinants are dominance and personal relationships. Everyone in the group has a perceived position in the dominance structure and, by forging relationships, can influence their relative dominance position. Life may be short and brutish, but it was logical, understandable, and workable. Civilization changed all that.

With civilization came wealth, and soon control of wealth became a way of asserting dominance. Wealth originally was simply the control of the grains harvested at specific times of the year and stored for use when there was no grain to harvest. Wealth was control of the flocks of sheep and goats kept for milk, fleece, and meat. An individual with more animals had more clout than an individual with fewer animals, regardless of physical strength.

Intelligence, guile, and cunningness were difficult, if not impossible, to project beyond the group in hunter-gatherer societies. But in the more settled permanent groupings of farmer-herders, such talents could be put to good use in the pursuit of status and dominance. An individual with many animals but little physical capability could pay another individual, who is stronger but poorer, to do his fighting for him. Such an individual could even use these employees to enlarge his holdings, thus enabling the dominant individual to expand his influence.

Here we see the first stirrings of dictatorship, war, economics, and civilization. We also see the beginnings of the extension of dominance beyond the group. The history of civilization is a history of dictatorships, war, and the buildup of wealth for the purpose of extending dominance or as a prize to be taken by those intent on dominance.

We are social animals. We are dominance-driven animals, some of us. And it is central to our understanding of human nature that the dominance equation be considered. We share a predisposition toward dominance with many other social animals. But we are different, too, not so much behaviorally as physically. Before the consequences of our dominance propensity are developed, perhaps a short digression into our evolutionary past is in order.

CHAPTER 6

We do not look like other animals. We are different. We use two legs rather than four, we don't have fur to any extent anywhere except on top of our heads and at the junction of our arms and legs, and we have no natural weapons. We have no claws, no fangs, and nothing else that could be used as a survival tool. Under these circumstances, it is easy to slip into the delusion that we are somehow, in some way, not animals but something better, different, even important.

It is true that we look like no other animal on Earth, not even close, but we are no different as to evolutionary history. We, like every other life-form, evolved, and how that process occurred haunts us to this day.

Present anthropological theory opines that we went from a quadrupedal stance to a bipedal stance on the plains of Africa. This contention is almost certainly a delusion. Consensus states that in order for us to see better or reach higher or for some other unknown and possibly unknowable reason, we evolved from an animal much like a present-day chimp to a creature much like ourselves. Further that this evolution took place on the veld of Africa about six to seven million years ago. The consensus does not address our loss of body hair or of our continuously growing head hair or our diving reflex or the web between our thumb and fingers or subcutaneous fat, but then, it is only a theory.

All the present theories of how we came to be bipedal with one exception also suffer from the "so that we could" syndrome. We became bipedal "so that we could" see farther on the plains of Africa. We developed an upright stance "so that we could" use our hands to manipulate tools. We walk on two legs "so that we could" display our sexual equipment and better entice mates. Man became an erect primate "so that we could" more easily shed the heat of the African sun. This last theory also points to our lack of thick body hair as an additional sign of the sagacity of this concept. The theory said nothing about cold African nights or the layer of subcutaneous fat we all sport mostly to keep warm. All these theories share one nagging problem. Evolution doesn't work in the "so that we could" mode.

Evolution, as we have seen, works in the here-and-now mode. If it works right now, if it helps existing individuals survive and procreate, it is selected for. "So that we could" doesn't work.

Take tool use for instance. How would the selection process work? Think in terms of a chimp; they use tools, they even make very simple tools, so we know that some of the wiring in their brains is conducive to toolmaking and using. But what would be the next logical step?

Present-day chimps live in forests and for good reason. Predation is minimalized when safety is close at hand. What if we took these same chimps and placed them in a savanna, an open grassland with scattering of trees? If we add to this scenario a predisposition on the part of the chimp to use tools, how do we get a troop of chimps to go from a quadrupedal gait to a bipedal gait? How would the selection process work?

It is difficult to postulate a scenario that would find any advantage in the very first step toward upright stance coupled with tool using. Especially a selective force strong enough to cause an animal to make the transition from a quadrupedal stance to a bipedal stance.

We would have to start with bone and wood tools as these would be most available. Horn cores are a good example. The bony core of antelope horn is fairly hard solid bone and naturally comes to a point. Wood stems burned in a grass fire also tend to form hard points especially if the wood was green when it was burned.

Okay, so our chimp band has taken advantage of this natural weapon/tool windfall, and they are searching the savanna for food. They use the horn cores and hardened wood sticks to dig up bulbs, shoots, roots, the occasional grub, lizards, insects, and club the random rabbit or newborn gazelle. This is coupled with fruit from the few remaining trees on the veld. Our chimps periodically stand as upright as they can to survey the plains for predators. Those that stand taller and see farther improve their survival, so we can say that the selective process would be for individuals who stood upright on their hind legs if even only momentarily.

But what happens when predators are spotted? Well, the alarm is sounded, and all our chimps throw down their tools and make for the nearest tree. All except one who decides to hold on to his tools and proceed tree-wise on his hind legs. He is eaten.

Let's say that through some perverse proclivity of nature, our chimp troop begins to become more and more bipedal. At some point, they will be approximately halfway between horizontal and vertical. Now what happens? They can no longer run very fast on four legs because they are too upright. At the same time, they can do little better than a stumbling swagger in the bipedal mode. When predators show up, all are eaten. There has to be a better way, and there is.

Before we proceed, a disclaimer: The following idea is not original by the author. It was first enunciated by Allister Hardy and later popularized by Elaine Morgan in her book *The Descent of Woman*. It is logical, it is reasonable, and it is possible. However, it was not the idea of an anthropologist, and so it just cannot be right. Most anthropologists I talk to say that it is a needless complication. On the other hand, both Morgan and Hardy can take solace in the fact that anthropology has been singularly unsuccessful in poking even the tiniest hole in the theory.

We started out as just another ape. We were probably a bit smaller than present-day chimps but not by much. Our home was the east coast of Africa. We were creatures of the woodlands with one difference from other woodland apes. We had learned that there was some advantage in living next to the ocean besides the view. At times, we foraged in the shallows eating fish trapped in tidal pools, clams, mussels, and anything else available.

This is not so far-fetched a scenario as might first appear. Borneo is the home of the proboscis monkey. This is a primate that is primarily arboreal but has been known to venture into the mangrove shallows to move from place to place and to forage. Our ancestors were, I think, a bit larger and a bit more at home in the water but not by much.

But how then did we become bipedal? This is an almost unbelievable change from our primate ancestors. True, chimps can hold themselves upright and even take a few stumbling steps when their arms are full of grapefruit or bananas. But when a chimp needs speed, they are on all fours, and they are fast, whether chasing a rival or fleeing a leopard. So while chimps can provide us with a possible pre-bipedal example, every bit of evidence suggests that the way to go, if you really want to go, is on all fours. And yet we, close cousins to chimps, are bipedal.

Why do we have so little body hair? And why do we have this weird hair-growth pattern? You know, head hair. Our hair and growth thereof, and lack thereof, is unparalleled in the animal world. Elephants, hippos, and rhinos are more or less hairless although all have hairs here and there. But these animals are big. So are whales for that matter. And that may have something to do with it. Naked mole rats are just that, but they live entirely underground, and this may have some impact on their hairlessness. What impacted our hairiness?

It is not just hairiness. We are the only species that has continuously growing hair at one place and one place only on our bodies. This is not a peculiarity of our genetic makeup either. We do have body hair, more or less. In spots, it is quite thick, and at the same time, it cycles. That is, it grows for a period of time and then falls out. Not so our head hair. It grows and grows and grows, and there has to be a reason.

Why is our nose shaped the way it is? Most people don't think much about their noses unless it's too big or pushed off to one side. But the shape of our nose is different from all other mammals with one exception. We share a projecting proboscis with the proboscis monkey, and the reason why shall shortly become clear.

Why do we have a layer of fat just under our skin? Some of us have more, some have less, but we all have it. It is called subcutaneous fat. Among primates, it is an exclusively human condition.

But bipedalism is the real question. Why are we bipedal? It is a condition so far removed from every other primate to be unbelievable, and yet here we are. And from the evidence of our past, we have been this way for nigh on to four million years or more. Long before we were big brained, we were long-legged.

It doesn't make any sense. There is no logical reason to be two legged as opposed to four legged. You cannot run any faster. Both chimps and baboons can scoot along at a real good clip on all fours, certainly faster than we could on two. As for lions, leopards, hyenas, and cheetahs, we were no match by any means for their speeds. So flight wasn't the answer. But on the other hand, to flee or to be able to flee would have been essential. We had no other defense, no claws, no fangs, a little bitty brain, and only two legs. The answer lies in loser evolution.

As I have said, our ancestors were apes. They, at one time, looked much as chimpanzees do today. They also lived much the same kind of life as chimps do today. Our ancestors occupied woodlands, living on fruit, leaves, insects, and, on occasion, monkey meat.

Groups of chimps consist of a number of adult males, usually more females, and offspring of various ages. Male chimps establish dominance hierarchies often with one male seeking out another male and forming an alliance through grooming behavior and socializing.

These partnerships then dominate the group. This has been success-ful behavior for chimps for hundreds of thousands of years and, if it weren't for man, would still be successful behavior today.

We no longer look like chimps. The change is one of enormous magnitude and is responsible for the contention that we are some-how different, better, somehow more advanced. In short, a superior "bipedal" animal.

There are additional differences that tend to make us think of ourselves as superior. We are *Homo sapiens sapiens*. Loosely trans-lated, that means "man, twice wise." Needless to say, we named ourselves and were unfailingly flattering in the process. But we do have a certain amount of intelligence, and since intelligence is an expensive commodity, there has to be a good, sound evolutionary reason for its development. Just like intelligence, our bipedal gait, continuously growing head hair, and a host of other characteristics are also evolutionarily shaped. This is the story of that evolution.

Once upon a time on the shore of the Indian Ocean on the east coast of Africa lived a band of chimp-like animals. The time is seven to eight million years ago. But other than that, things are pretty much the same as exists today in Africa. The shoreline where our chimps live consists of sandy beaches broken up by small rocky out-crops and coral-laced tidal zones. It is a land of golden yellow sand, black rock, and blue-green sea. The beach is about twenty to thirty yards wide, mostly wave-washed sand interspersed with piles of rock and rubble. The beach borders on a small bay. The bay is almost three miles wide and extends out into the ocean just under a mile. The arms of rock forming the bay curve in an incomplete semicircle, affording two or three gaps on the seaward side for the tides to enter and leave. As a result, the little lagoon tends to be relatively calm even at those periods when violent storms rage in the ocean beyond.

Behind the beach, the land slopes very gradually to a series of small hills running parallel to the shore. These hills are covered with trees, bushes, and vines. Approximately four miles west of the beach, a small river meanders through a gap in the hills, turns south for a couple of miles before finding its way through the remaining hills and into the southern end of the bay.

The chimps live in the area bounded by the river and the ocean. This is their territory, and it is defended more or less vigorously depending on troop strength. Fortunately, this group occupies a relatively rich niche. Overall strength is generally high and has been that way for many millennia. The reasons for these conditions are many. The hills bounded by the river are rich in fruit for most of the year. Figs, especially, are common, and the troop has a collective memory of when each tree ripens. There are several kinds of monkeys that occupy the chimp range, and the males have become adept at trapping individual animals against the river. Monkey meat makes up a major portion of the small but significant protein element to the chimp's diet. Another source of necessary protein is the seashore. Small crabs crouching under rocks at low tide make a welcome addition to the mostly vegetable and fruit diet of the animals. The big disadvantage of the seashore is that the chimps do not like to get their fur wet.

All in all, the group has it pretty good. The area they occupy provides for all their needs, and for the most part, the chimps prosper. It was, however, not all Eden. Once in a while, one of the youngsters would fall out of a tree and kill itself. A pit viper, unintentionally disturbed, would strike, and a chimp would die. Leopards also patrolled the area but not very often. The trees and brush prevented the grasses, so beloved by antelope, from growing, so leopards found generally slim pickings in the chimp's territory. It was a pretty good life for a chimp, but that was about to change.

Over a period of several thousand years, the local weather patterns changed. Monsoons—long a staple to water the trees and bushes—failed. At first only once in ten to twenty years, and the chimps hardly noticed. In fact, those years when there was no rain were happy times for the animals because they disliked the downpours.

The reduced rainfall did have one serious consequence, but it was one that the chimps were totally unaware of. The water table began to drop. This did not affect the river because its water came from rainfall far to the west, but locally, bushes and vines started dying as the rains failed and their roots were unable to tap into the groundwater. Even more ominous, no new bushes and vines were sprouting. Or if they did germinate, they were dying before they could get established.

Open areas started to appear in the chimps' territory. Large trees would die, topple to the ground, and not be replaced. The twenty-four to thirty inches of annual rainfall dropped to just twelve. Not enough for trees but enough for grasses.

And so the grasses came, filling up the openings in the forest, creeping down from the higher hills where water was less available and into the valley. Soon the chimps' territory began to resemble open woodland/savanna. The presence of the grasses brought the grazers, which in turn brought the carnivores. The small, snug, safe harbor of the chimps was changing, and these changes, occurring over several thousand years, were to lead directly to us.

The change was slow but inexorable. The first to go were the monkeys. Trees became so far apart that the monkeys were unable to get from one to another without exposing themselves. Eagles and hawks took more and more of the youngsters until the animals could no longer form viable troops. As the gaps between the trees got wider, traveling became more and more hazardous, until finally, there were no more monkeys left.

Without the monkeys as a protein source, the chimps began to rely more and more on the little white crabs that spent low tide under the rocks. They still did not like getting their fur wet, but a hungry belly trumps wet fur every time, and gradually the chimps came to depend more and more on the ocean for food. Fish were often stranded in tidal pools. One chimp found that pounding a stone on some of the shells fastened to rocks along the shore sometimes produced small bits and pieces of mussels. Carrion, too, sometimes floated into the bay on the tide and was quickly eaten by the chimps.

Our chimps could have moved. There were better areas beyond the river. But these areas were occupied by bands of chimps larger and stronger than our bayside tribe. Our little group could not compete. They had become loser chimps.

By now, our one-time robust troop had been reduced to three males, five females, and a few offspring. Birthing was uncommon in these lean times, and females sometimes had problems providing enough milk for their babies. Leaves and now grass seed heads as well as insects made up part of the chimps' diet, but their caloric needs were barely being met, and chimp health began to suffer.

Things were about to get a lot worse. One day the chimps were foraging along the beach at high tide when six lionesses appeared from around a small headland about one hundred yards up the beach. Intent on feeding hungry bellies, the chimps, at first, did not see the animals. The lionesses watched the chimps intently for a couple of minutes. They had never hunted chimps before and were not sure if they would be suitable prey.

Then one of the younger chimps spotted the cats and screamed. All eyes turned in the direction the screamer was looking, and the whole troop started screaming and hooting. The chimps were at least sixty yards from the nearest tree, and there was no other security on the whole stretch of beach.

One of the males started for the trees as soon as he saw and recognized the danger. The male was fast, but two of the lionesses immediately started to cut him off. The cats still weren't sure if this was food, but moving animals triggered a chase reflex, and in twenty-foot bounds the lionesses were after the chimp. They quickly covered the ground between them and caught the male ten yards short of sanctuary. As the chimp turned to defend itself, the first cat hit him with her shoulder, sending him head over heels into a dazed sprawl facedown in the grass. Before the chimp could gather his wits the second lioness was on top of him, and a single bite through the back of the neck into the skull killed him.

The chimps on the beach screamed again and again as the remaining lionesses started down the beach toward the tiny group. There was nowhere for the animals to go. The tree was out of the question, and there was nothing else except that about forty feet out from the shore was a small pile of rocks with one rather large roundish boulder about ten feet high right in the middle of the pile. The chimps had already been out to the island but only at low tide when only two or three inches of water covered the space in between. Now the water was twenty-five to thirty inches deep, and none of the animals had ever been in water that deep before.

With much hesitation and screaming, the chimps backed into the water as the lionesses approached. Hooting and splashing, the chimps reluctantly waded across the gap and climbed up on the boulder. Wet and bedraggled, they sought tiny fingerholds on the rock and climbed up to its top. They then turned to face the lionesses. Waving their arms, hooting and screaming, the small band tried to put up as brave a front as possible. The two remaining adult males stood at the front edge of the boulder, ready to defend the troop as best they could.

But these lions liked water even less than the chimps, and the apes watched in wonder as the animals paced back and forth at the edge of the beach. One would approach the water, eyeing the distance as if to make a single bound to reach the island, only to back off when a wavelet washed over her paws. Before long, the big cats lost interest in the chimps on the island and went over to where the other pride members were dining on the male.

The six lionesses made short work of the chimp and, still hungry, returned to the shore nearest the small island. Some again tried to enter the water, only to retreat when the top of their paws got wet. The chimps watched the big cats, all the while hooting and screaming every time one started to enter the water. After a time, some unseen but recognized signal passed among the cats. They left off trying for the remaining chimps and, with a backward glance or two, resumed their stroll down the beach. They stopped at the stream and drank, took one last look at the chimps stranded on the rock, and padded off up the river.

The chimps continued to hoot and scream but with less and less urgency. They had had a terrible fright, and now much time was given to pant hoots of reassurance and pats of sympathy. Slowly the troop quieted down, and one of the males started down off the rock. The females would have none of it. With much hooting and panting, they convinced the chimp to return. Night was fast approaching. The cats might be hiding just over the hill, and all the females were still terrified. The chimps spent the night on the rock. The first time in hundreds of thousands of years that the animals had not spent the night in a tree.

And that's how it started. That's how we came to be us. All because lions do not like to get their feet wet. The chimps began to spend more and more time in the water. They found that farther out in the bay, the pickings were even better. Clams were to be found in the sand, as well as scallops. But the chimps needed to refine their mussel-breaking skills on these new sources of food. Before, it was just a matter of smashing a rock against the shells attached to the rock. Now they had to forage for clams, collect them, bring them to the rocks, and then smash them open. The animals were up to it though and, over a period of less than ten thousand years, became proficient enough at acquiring sustenance from the sea that they almost entirely quit the land. Their only daily landward excursion was for drinking water from the stream. They had become aquatic apes.

This new environment put a different selective pressure on the chimps. They gradually got over their dislike of getting wet and even began to get some pleasure from splashing around in the shallows. All this splashing and wading did cause one problem, however. The chimps' fur became wet and matted on a regular basis. This caused problems for some chimps. Cold and wet, their health was compromised, and they did not do well. Selective pressure was for chimps with less body hair. They dried quicker and were less likely to have problems than their hairier brothers. Over a period of time, the chimps lost most of their body hair, retaining only fragments here and there.

It could just as easily have gone the other way. Selection could have been for a fine waterproof fur coat. Nature doesn't care. If naked works, if naked and subcutaneous fat works, use it. It is possible that our body hair, then like chimp hair today, was so coarse that it was easier to lose it than use it. And so we did. But what about head hair?

Head hair grows and grows and grows. Lengths of ten to twelve feet have been reported, and three-to-four-foot lengths are rather common. There is a reason for this. Baby chimps use the fur of their mothers to hang on. Most baby chimps cling to the back of their mother when she is traveling, using the fur on her back to hold on. But our aquatic apes were losing their fur. In the water, the babies rode high up on the female's shoulders. They needed something to hold on to. So while selective pressure was on to reduce body hair, it was also on to produce hair in a certain area where it could be used by the youngsters.

But the loss of hair resulted in another problem. Seaside nights are cold even in the best of weather, and without body hair, our apes suffered. Our ancestors were probably already black. Living at the edge of the ocean exacerbates sunburn, and selection would have been for those individuals who had greater amounts of melanin in their skin. This color would also have been beneficial on cool mornings when the dark color would absorb the warmth of the early morning sun.

But cold was still a problem, and the answer was subcutaneous fat. Individuals with small amounts of subcutaneous fat improved their chances of survival. Gradually, the species came to have that really ugly-looking translucent grayish-yellow fat just under the skin that is unique to our species.

Other changes, too, were brought about by selective pressure. Even before they became aquatic, the apes could walk on their hind legs, not very well, not very far, but they could do so, just as modern chimps can. In the water, this ability became even more valuable. The farther out into the bay the apes ranged, the better the pickings. But to get to the best areas, they would need to wade through water two and three feet deep. Selection was for animals that could stand more upright. Each incremental genetic predisposition toward a more upright stance was selected for. Any genetic advantage that would permit greater aquatic mobility would tend to survive at the expense of less adaptable animals. In the end, selection was for animals that could stand fully upright.

In deeper waters, it became necessary at times for the chimps to actually swim. Some of the chimps ended up with serious sinus infections from water forced up their exposed nostrils. Over a period of time, a small projection of skin and cartilage just above the nostrils grew as it was selected for. Today, we call it a nose. And it still helps us when we dive into the water headfirst by protecting our nostrils. This is evolution in action.

But evolution is not always the improvement we would like to think it is. Evolution is the compromise that organisms make to survive. One compromise we made long ago in that forgotten Indian Ocean lagoon still haunts us today.

Because apes that could wade into deeper and deeper water were selected for, our bodies gradually became upright. This is how evolution made us bipedal. Each incremental shift toward the vertical was rewarded by enlarging our feeding area. This is how evolution works, rewarding the here and now, not some "so that we could" future.

Slowly our bodies changed. In the safety of the water, free from predation, we made the complex adjustment from a horizontal, quadrupedal body to a vertical, bipedal figure. And we paid a price for that change.

The difference between a chimp pelvis and a human pelvis is substantial. The human pelvis is a shallow bowl, a bit wider than deep with a small hole in it. A chimp pelvis, on the other hand, is much deeper than it is wide and has a good-sized hole in it. The key is not the shape and size of the pelvis but rather the size of the hole.

As our pelvis rotated and shrank to accommodate our new upright posture, the hole also shrank. This opening houses the birth canal, and its size has pitted our expanding brain against the ability of the pelvis to provide birthing room ever since.

It was not too much of a concern for our ape ancestors. If it had been, we wouldn't be here. Their brains, at birth, were small, a little bigger than present-day chimp brains at birth. The problem was not so much brain size as it was head shape. It's a matter of muzzles.

All mammals have muzzles. All, that is, except us. Muzzles are important, necessary even. Muzzles are essential to both herbivores and carnivores. They are handy tools for feeding and fighting. They multiply the power of the teeth by applying the force-times-distance equation. Teeth at a distance supported by strong jaw muscles will result in a strong clamping force. Just ask Crocodile Dundee. The concept also works for herbivores. Teeth at the end of a muzzle can be used to pull up grass, nip off shrubbery, and pick up stems and twigs efficiently and cleanly. Muzzles are critical to most animals, a standard natural issue because they work.

On the isle of Beata, off the coast of the Dominican Republic, there lives *Sphaerodactylus ariasae*. *S. ariasae* is a gecko, a very small gecko, a teeny tiny gecko. This gecko is shorter than its name. At one and a half inches, including tail, this gecko is the smallest reptile in the world. A gecko so small that a flea, or maybe two or three, would be a meal. *S. ariasae* has minute toes and arm bones the size of the letter *i* on the dime. The animal also has a muzzle. Tiny, true, but a chomper nonetheless—with teeth and tongue the better to bite you with.

Then there is *Hippopotamus amphibius*. Its whole head is a muzzle. With huge tushes and wide fat lips, the hippo is the epitome of a muzzle. Then there is everything in between. Horses, antelope, cats, dogs, you name it, they have it—a muzzle. Muzzles are important, yet we do not have one. Why?

All apes have muzzles. Chimps have muzzles with imposing canine teeth that are used for threat display, fighting, and engulfing grapefruit. Our aquatic ancestors certainly had muzzles. But we do not, and fossils of our upright ancestors show that they did not have muzzles either. Evolution is the compromise that organisms make to survive. And this was one of the most profound compromises any organism was ever likely to make.

Head shape is the key. Measured from the end of muzzle to the back of skull, our forebearers' heads, while not very wide, were long. This length had to pass through the pelvic opening, which was shrinking as we became more and more upright. Selection was for babies with shorter and shorter muzzles, until at some point muzzle length was so small that it no longer had any value as a survival mechanism. With the shrinking muzzle came the shrinking canines. Without the muzzle length to allow the gaping to provide leverage, selective pressure on large stabbing canines was reduced to a point where they became useless.

We lost our canine teeth, but we gained a brain. More correctly, a better brain than our ancestors had. The water we lived in favored individuals that could stand upright. Upright stance changed the shape of our pelvis. Changing pelvis shape selected for heads that were rounder than long from front to back. Heads long from top to bottom were not a problem, and so head shape changed and so did brain shape. Areas of the brain, once small and unimportant, became larger and capable of taking on functions that a primate with no canines and no claws would find absolutely essential to survival. Our brain did not change to make us smarter. It changed to ease our birth, and it was only afterward that the selective process began for a larger and larger brain. Our contention that intelligence is inevitable, or essential, is a delusion. The first step toward our present brainpower was simply the by-product of our need to get through a smaller birth canal. It was only later that we were able to use our accidentally acquired brainpower to survive.

Survival is, at best, a chancy thing, and over the millennia, survival patterns have fallen into several broad types. Some organisms opt for armor. Turtles and armadillos come to mind. Some come equipped with camouflage. To blend in is not to be seen; to be missed is to live. Sloths are slow but well blended. They are one of the most successful animals of the Amazon basin. Their survival depends on camouflage. Elephants are huge and have, accordingly, few enemies save for man. The downside of elephant survival strategy is that they must eat almost continuously in order to sustain their evolutionary survival strategy. And so it goes, animals that survive have successful coping strategies.

Homo is the only organism forced to rely on brainpower for survival. It worked, but it is not a common survival strategy, and we are the only animals that have ever succeeded at it.

No claws, no fangs, slow, small, with only a brain accidentally shaped by a revolving pelvis with which to survive, it's a wonder that we did. In the process, we took advantage of the one attribute available to us and our evolving brain. We survived by learning in ways and by processes available to no other animal.

Most animals must learn in order to survive. Very few are hard-wired to a point where all behavior is automatic. Mammals are especially adept at combining learning and instinct to survive. Primates show considerable skill at not only learning, but also at being able to teach others by example. This is our rather tenuous survival mechanism. Using the only tool we had—our brain—we survived.

It started slowly. Remember the mussels that adorned the rocks that we bashed with small rocks and, in the process, produced small amounts of protein for food. Over time, with our differently shaped brain, we became more adept at mussel bashing. We chose better and better tools, adding clams and scallops to our menu. But that was just a start; small fish in tidal pools could be caught by hand, but the process was tricky. Water reflects light, and fish, even in shallow water, are not where they appear to be. We learned that if the grab was made here rather than there, fish became dinner.

Then someone found a piece of wood—long and tapered, smooth from rolling in the surf, and just the right size. Now the little fish in the tidal pools were dispatched with ease, and even bigger fish could be struck provided that one waited quietly and patiently next to one of the numerous channels that drained the tidal pools.

Over several thousands of years, we survived and changed and became more adept at surviving. This allowed more change to occur. And tens of thousands of years after those nasty cats chased us into the ocean, we emerged upright, with stones and clubs and, most of all, with the ability to learn in ways not afforded to our chimp-like ancestors. We were well on our way to becoming *Homo*.

There is yet one other aspect of this survival by learning that must be taken into consideration. But before that, a small digression. It is 1953, and the sweet potato monkeys are at it again. Specifically Imo. Imo is a one-and-a-half-year-old Japanese macaque. Her home is the island of Koshima off the southern coast of Japan. Japanese macaques live in the forest on the island, and in 1953, with primatology just coming into focus, our island macaques presented a possible place to study monkey behavior.

There was a problem. The monkeys lived in the forest on the island, and suitable methods for the study of primates in place had not yet been developed. The answer was to entice the monkeys out onto the beaches of the island. And that enticement was sweet potatoes. The primatologists put out sweet potatoes, and the Japanese macaques responded by coming down to the oceanfront for a daily picnic.

And so the macaques came, and the primatologists studied and took notes and gauged social relationships. And an excellent time was had by all, except when the sweet potatoes were placed on the beach and they picked up a nice coating of sand. For a while, the monkeys made due with a bad situation, but then one day, Imo solved the problem. She took her sweet potato over to the ocean and washed off the sand. Soon, Imo's playmates, mother, and older sisters were washing their sweet potatoes. Imo had discovered a new way to do things and was able to teach, by example, something that would benefit the entire troop.

This is the advantage of brainpower. Was Imo a genius monkey? Certainly, she demonstrated a grasp of problem-solving techniques that point in that direction. Some researchers say that this behavior was pure accident. But even if it was, we must still acknowledge that Imo realized that this purely accidental action had the desired effect of producing a more edible sweet potato.

Learning is the only survival mechanism that can be passed on in this way. Once our brains—shaped by accident—became adept at learning, the whole evolutionary process was accelerated at an ever-increasing rate. Selection was for individuals who could learn fast and who could absorb more and more complex behavior patterns. And unlike some survival mechanisms, there is little or no limit to learning. We were on our way to becoming *Homo*, and the occasional genius invented and taught and made survival that much easier.

One of the aspects of intelligence that does distinguish us from all other animals is the occasional genius that would arise in a given social unit. The fire bow was invented—I am sure—by a prehistoric Einstein, as was the atlatl and the hunting bow. These, as well as many other inventions and processes, then spread because while the run-of-the-mill hunter would not have thought of the idea, he was enough of a craftsman to make the invented item or implement the new process once someone else invented it. If your social unit has bows and arrows while your neighbor has not, you have a tremendous advantage in the competition for survival. This was the key, the true difference between our survival mechanism and all other survival mechanisms.

Every once in a while, a genius would be born, much smarter than the average protohominid, and behavior would change rather dramatically over a relatively short period of time. From now on, our brains would be selected for learning, for the ability to survive by knowing, because we had nothing else on which to base our survival.

It was a successful, if extremely uncommon, survival mechanism. And we prospered in the shallows of our little bay on the coast of the Indian Ocean. Over time, we outgrew the bay.

This ushered in a period of extreme stress for our protohominid ancestors. There were more than one hundred individuals now attempting to make due in the bay, and resources suffered. Food became hard to find, and population densities began to drop, and less robust individuals succumbed to the pressure of not enough food. Food fights began to occur, with stronger, more dominant individuals forcing less assertive animals from the few remaining food sources. There was trouble in paradise, and natural selection again would lend a hand.

Evolution has always functioned thusly, and it is nature's way. We, in our human billions, are the only ones to violate the laws of evolution in this matter. Under natural law, species find a balance with respect to their niche. The process is curiously simple. A species will exploit a given habitat. It will grow and expand until it reaches the limit of that habitat.

At this point, control mechanisms kick in, not conscience-thinking mechanisms, but rather natural ones. One of the most common population-control mechanisms is that of overpopulation. When a species breeds under favorable conditions, it sometimes gets to a point where it exceeds the carrying capacity of the habitat. At this point, none of the animals get sufficient food, and the population crashes. Only the strongest, healthiest animals survive to begin the process all over again.

Another control mechanism is predation. The animals become numerous enough that predators move into the habitat in large numbers to harvest the population. This can go on until the species is hunted virtually to extinction or, more likely, until enough have been hunted that the remaining animals no longer constitute an easy food source. In one way or another, natural selection sees to it that a species maintains a balance within its habitat.

In the case of our ancestors, a third possibility presented itself. With our budding intelligence, we had the ability to exploit different habitats. Not much different, but different nonetheless. North of the bay, the yellow sand beach stretched for nine or ten miles before coming to an end against a headland that extended more than three miles out into the ocean. There were a few islands offshore from this beach as well as coral reefs that kept large waves from breaking on the beach. This habitat was not as good as the bay, but it could be exploited. And it was.

From one small bay into which one small band of chimp-like primates were forced, there emerged a bipedal, long-haired, aquatic ape capable of finding both shelter and sustenance in the surf. We no longer had large canines, but we did have both stone and wooden tools as well as the brainpower to use them, brainpower which resulted from the selection process involved with becoming bipedal. We also benefited from the occasional individual with greater intellect. Most of us were a little better-off mentally than chimps, but we could learn and retain what the intermittent genius thought up. We were successful, we survived, we grew, and soon, probably in fifty thousand to sixty thousand years, we occupied every suitable nook and cranny of shoreline on the east coast of Africa. The hominids had arrived.

CHAPTER 7

The coast of any body of water is a desirable place. Houses situated on coasts or lakeshores or riverbanks are always more valuable than houses situated inland away from the water. But there is a problem with such desirable spots. While there are literally millions and millions of square miles of oceans and millions of square miles of land, the land-ocean interface is a very limited habitat. This habitat is further limited by conditions. Where cliffs descend into the sea, where broad sand beaches extend hundreds of yards out into the ocean, preventing sea life from getting started, or even areas where surf and tide are so strong that little or nothing of use is to be found, creatures like us could not survive.

This was the problem we now faced. We once occupied a bay and found it overcrowded. Now we occupied a whole coast, and it, too, became overcrowded. We had two choices. We could enter the sea and become aquatic in the manner of seals and walruses, or we could go back onto the land from whence we came. We may have tried both, but only one worked, and that was the land.

One of the inadvertent results of our changing brain shape was the ability to adapt. This, more than anything else, characterizes the human species, and we developed this capability in the water off the east coast of Africa. Like all habitats, our oceanfront home varied; some places like the bay were close to perfect. It had a good supply of food, numerous islands suitable for refuge, and surrounding arms of land to soften the wrath of the ocean. All in all, an original Eden.

But a lot of the coast was not Edenic. We adapted. We learned how to forage, how to time tides, how to judge waves, and how to be on hand when the spawning runs occurred. We learned how to dive into the surf, swim underwater, and come out beyond the surge. The ocean taught us many things, and with our new brainpower, we learned and prospered and multiplied.

Soon all the bays were occupied, the open beaches populated, and most desirable sites with both food and sanctuary taken, over-populated, and still we grew. Some of our ancestors in less desirable coastal sites started using resources available on land. We went back to the trees from which we came.

Five million years ago, our ancestors came out of the water. Not because we wanted to, but because there was no more room on the seashore and because we were unable to go into the sea full-time. This is loser evolution in action. At first, the land served only as a place to gather food, insects, roots, and small animals. Where the sea could not supply food but could offer safety, our protohominid progenitors made the first tentative steps back from the ocean.

Most of the attempts were unsuccessful. True, we had wooden and stone tools and weapons, but we could not compete with animals already in place in the open woodlands that abutted the ocean. And so one attempt after another failed. Small groups would establish themselves or try to, and predation would outpace reproduction. Overcrowding would occur on the beach, and another group would be forced to try their luck on land, only to lose the battle and the war.

Then one group gave birth to a genius and survived. The idea was quite simple. The sea had been our refuge. It protected us from predator attacks, and what we needed was another refuge. We went back to the trees from which we originally came. Everything heretofore has been speculation—informed, reasonable speculation, but speculation nonetheless. That our bipedal ancestors were tree dwellers is not speculation. And for that fact, we can thank Lucy.

There were probably others before her, but Lucy is the oldest hominid that is complete enough that we can make some judgments about how she lived. Lucy is just over three million years old. She is, or was, about three feet, eight inches tall, weighed about sixty-five pounds, and had a brain about one-third the size of ours. Her skeleton is 40 percent complete. She had an apelike face without the muzzle, and had a bipedal gait. I would be willing to bet that when she was alive, she was hairless, black, and had long matted head hair in what we would refer to today as dreadlocks. The first aquatic ape to successfully reconquer the land of her ancestral apes. At least, she is the first protohominid that we know anything about.

And one of the things we know is that her arms were proportionally longer than ours, and her finger bones had a slight curve to them. Both of these characteristics would have been helpful to a bipedal ape attempting to conquer the land by taking to the trees.

This makes sense. A prerequisite for procreation is survival. Every prey animal needs a survival mechanism. It is something that anthropologists sometimes forget. In the water, we were safe because land predators could not or would not come into the sea to prey upon us. At the same time, we were a relatively new introduction into the ocean, and marine predators had not yet zeroed in on our species.

On land, we were preyed upon by the ancestors of the present-day predators of the African veld. We had very little in the way of defense. We had sticks and stones only. We had no canines to speak of, no claws. We should have been quickly wiped out, except we could climb trees. This saved us.

Hyenas do not climb trees. They are not built for climbing. Lions can climb trees but are not very good at it and are reluctant to climb even a small tree to get at prey. Leopards do climb trees and regularly stash their kills in trees, so leopards would have been a problem, but a manageable one. When the spotted cat climbs a tree, it loses two of the three characteristics it uses to dispatch prey. The leopard is fast, it has long sharp claws, and it has long sharp canines. In the trees, especially on smaller branches, the leopard loses its advantage of speed. It must move slowly or risk falling from the tree. It also loses the use of its claws as a killing mechanism. The cat must hold on with its claws and their usefulness as a weapon is substantially reduced. Baboons learned this a long time ago and today find the trees a reasonably safe refuge from leopards.

Lucy and her kind prospered. They spread over the woodlands of East Africa and survived for more than a million years—success by any measurement. And yet while lions and leopards are still among us, Lucy is not. And the reason is back in the water along the eastern coast of Africa.

We were successful aquatic apes, the smartest ones still in the water—safe, secure, and begetting. Populations grew, pressure increased, and natural selection selected. There just wasn't a lot that nature had to work with. Because of our physiology, most natural survival mechanisms were unavailable. We were bipedal, so speed was out of the question. Because of the shape of our head and lack of gape, teeth—that is, large sharp teeth—were unavailable. And claws? To go from nails to claws was going to be difficult even though way back, our remote ancestors had them. By all rights, we should have gone extinct.

At this point, we had one minor advantage. As we went from a quadrupedal to a bipedal posture, our head shape was altered to allow passage through the birth canal. This alteration, even though it was not intended to, caused different areas of the brain to expand. The main area of expansion was to the frontal lobes, the site of cognition. Our success was to depend on our brain. The ability to think, no matter how crudely, the power to reason on a very basic level—this was all we had, and in the end, over millions of years, it was to prevail. Even though our mental abilities were an accident of birthing, there is no doubt that without this capability, we would have, almost certainly, gone extinct.

One day the bay troop awoke from its island refuge to find a dead dolphin washed up on the beach. The animal had only recently died and was still kinda fresh. The troop had seen animals like this before and knew that they probably would get no value from the carcass. Dolphin skin was tough, and the troop had nothing with which to penetrate the animal's hide. They would have to wait for others such as lions or hyenas to open the body, then, with any luck, the apes would scavenge leftovers. The animal's eyes were scooped out and eaten by the first male on the scene. The blowhole was poked and prodded, but it was obvious to all that there was no way to get at all that nice juicy meat. No way to eat that rich blubber, that four hundred pounds of dolphin that all the apes could have feasted on for several days. The troop set about looking for more accessible food, an activity that was occupying more and more of their waking hours as numbers increased and food supplies decreased.

One of the apes, a young female, had developed a feeding mechanism that allowed her to satisfy her hunger pangs fairly easily. All the apes used stones to pound mussels and limpets off the rocks. But thousands of years of pounding had resulted in very few shellfish in the tidal zone. Most of the mussels that were left were in a series of crevices in the rocks along the south end of the bay. The crevices were only six to eight inches wide, and using a rock to remove the shells was not possible. Although at one time or another, almost all the animals had tried to do so.

Our female was one who tried, only to fail and bark her knuckles in the process. Then a chance blow at the edge of the crevice loosened a dinner plate–size chunk of slatelike rock, one surface of which was covered with the succulent mussels. The young female had found a new way to feed, and she exploited it. Soon, she was joined by two other females. In the beginning, they only scavenged the remnants of the first female's meal. But eventually, one of the females also started to use stones to dislodge the slate on which so much seafood was growing. The troop was beginning to learn.

On the day the dolphin washed up on the beach, our female was at her favorite feeding spot using her favorite rock. When she struck a particularly hard blow, her pounding stone split in half.

One side was the base of a round fist-size pebble, and on the other, a roundish flake two inches across. It was roughly circular with a sharp edge on about three-fourths of the flake.

The female gave a small whimper of pain when the flake cut into two of the fingers she used to separate the broken rock. The female licked at the blood then looked at the flake. Her brain, a little bigger than a chimp's but differently organized, vaguely suggested an idea. The female left off, foraging among the shellfish crevices, and strode down the beach to the dead dolphin. Without hesitation, she squatted down and drew the sharp edge of the flake over the dead animal's skin.

As if by magic, the grayish hide parted, revealing white blubber underneath. Another stroke, another cut, and another line of blubber. The female cut through the fat to the meat below and finally to the ribs. The blade would go no further. But that was okay, the female cut off a chunk of blubber and began to devour it.

One of the females who had followed the blade wielder saw her companion eating and, in typical ape fashion, held out her hand in a begging gesture. The first female cut her a thick chunk of flesh, and for a short while, the two feasted in contented harmony.

Another female came by and begged and then another. Soon, a low-ranking male came over to see what all the fuss was about. When he saw one of the females chewing on dolphin blubber, he took it away from her and began to devour the treat himself. The female who lost her food screamed loudly and made a half-hearted attempt to threaten the male. He looked at her in utter amazement, all the while stuffing the blubber into his mouth as fast as he could.

Her cries brought the rest of the troop. The dominant males, seeing the others eating, shouldered their way in, scattering females and infants in all directions. They commandeered the body for themselves. The males were powerful animals, and once the opening up of the dolphin gave them a handhold, they speedily dissected the animal. Those who dined that day on dolphin were our direct ancestors, the very first animals to use modified stone tools.

And so our ancestors developed. The use of tools, especially sticks and rocks, had been around for half a million years. Some of our apish forebearers used rocks to smash nuts and twigs to fish for termites. They even modified the twigs by stripping them of their leaves in order to poke them into termite holes. But this was different. Not just any rock would do, not just any size either. The process called for more brainpower, and those with greater brainpower survived at the expense of less intelligent individuals. Selection was for smarter individuals because intelligence was the only trump card we had to play.

We could very easily have lost. Intelligence is a very poor survival mechanism. Speed is good, camouflage works, sharp claws and strong canine teeth do wonderfully well, and even size can be advantageous, but intelligence? We were the only animal that we know of to try it, and while in the long run it has been successful, in the beginning it was touch and go. For the longest time, it was a make-do solution that could very easily have failed.

Our brains were larger now. Toolmaking and tool using required greater manipulative capability, and individuals with that ability improved their survival potential. Manipulative ability required greater brainpower, and before too long, another ancestor was foraging in the shallows off the East African coast. We call him *Homo habilis* (handyman).

And handy he was, because of brainpower. *A. afarensis* had an average cranial capacity of 440 milliliters; *A. Africanus*, 450 milliliters; *A. Robustus*, around 500 milliliters; and *A. Boisei*, 515 milliliters. But *Homo habilis* had an average cranial capacity of 680 milliliters, and some individuals had as much as 800 milliliters. This is more than half our present-day range of about 1,200 to 1,500 milliliters.

But it was not just capacity; it was also the organization of that brain. Again, it was the way our skull was shaped that allowed the frontal lobe area to expand. These are the areas of cognitive thinking, the area of conceptual thought and reasoning. *H. habilis* is the first *Homo* that we know of that used stone tools.

Homo habilis still had a problem. The tools he developed and used made it easier to obtain food on the plains of Africa. But we still had no good way to deal with predation. Tools could be used to smash the bones of the carrion he scavenged, perhaps to dig roots or tubers, or even as a secondary implement. Short digging sticks would benefit from a sharp point. Oldowan hand axes would have done the trick nicely. So handyman, like *A. africanus*, found refuge in the trees. Perhaps he should be considered an *Australopithecus* in spite of his toolmaking ability. Perhaps *Homo* should be reserved for the next aquatic ape to take to the woodlands of Africa. An aquatic ape that was so successful, he came out of Africa and populated a good portion of Europe as well as most of Asia. Truly, a formidable achievement for the descendants of a small band of chimp-like creatures crouched in the shallows of some African lagoon in fear for their lives as lions stalked the shore.

The next hominoid to come out of the water was *Homo erectus*, the first to solve the problem of predation.

The common designation of an animal as hominoid is the making of stone tools. Perhaps a better definition of *Homo* might be the use of fire. *Homo habilis*, as far as we can tell, did not have access to or use of fire. *Homo erectus* did, and it was this ability that allowed *H. erectus* to conquer the world.

Not enough thought has been given to the concept of fire with respect to our ancestors and the impact this acquisition has had on our evolution. No animal in the world has ever adopted, as a survival mechanism, a chemical reaction that it could lose and not be able to reacquire. The potential for fire to drive our evolutionary journey is almost unlimited. There can be little doubt that we use fire long before we could make it. The selective process engendered by such a situation is what would lead directly to *Homo sapiens*.

Many years ago, Rudyard Kipling wrote a story about a boy named Mowgli. As a baby, Mowgli was lost in the jungle. The child was raised by wolves and came to think of himself as just another animal of the forest. More recently, Walt Disney made a movie of Kipling's story, calling it *The Jungle Book*. The movie may be animated, but its truths are real.

Early in the story, Mowgli is kidnapped by monkeys and taken to their leader, an orangutan named King Louie. The crux of the kidnapping is that King Louie would like to know, in some of the finest jazz lyrics around, man's "secret of fire." Alas, Louis is to be disappointed. Mowgli never came in contact with other humans. He knows no more about the secrets of fire than the wolves he was raised with. Louie wanted fire. Louie knew that fire would make him human. The secret of fire separates man from animals in the movie, just as it does in real life.

*****

A small group of aquatic apes (or should we by now say Homos?) occupy a stretch of beach fronting a large mixed wood-land/grassland plain. There are six males in the troop, ten females, and fifteen offspring of various ages. The males are about 5 feet 5 or 5 feet, 6 inches tall and, maybe, 125 pounds. The females stand at 4 feet 8 to 4 feet, 10 inches and weigh around 90 pounds. These *Homo erectus* are powerfully built and fully capable of running, climbing trees, and swimming in the water. They make their living mostly on land. The sea in their territory was almost entirely sand bottom, and little marine life grew in the area. The sea did have one advantage for our ancestors, however. About fifty yards offshore was a small sandy mound, which the troop used as a refuge at night.

These early humans lived by foraging and scavenging. They were equipped with stone tools including small fist-size rocks used as projectiles. They also carried short pointed sticks used for digging and for flipping over porcupines so that they could get at the soft underbelly with their flint cutters. These *H. erectus* flourished on insects, grubs, lizards, fruits, roots, and scavenged meats and the occasional porcupine. The troop prospered along their length of beach, and life was good.

The alpha male in the group was largely responsible for this prosperity. He was a genius. Through trial and error, he had discovered that some stones make better tools than others. Now most of the males had good, sharp flint hand axes. The other males also knew how to form the desired shape from raw stone by watching the alpha male. Even a few of the females carried flint tools, mostly cast-offs from the males, but a few made from scratch. These tools made life marginally easier for the whole troop. The male also made a discovery that contributed substantially to the success of this group of *H. erectus*.

Periodically, fire swept the plain, and in the course of burning, the fire would sometimes incinerate the base of a particular shrub. The shrub put out long single branches from a short gnarled base. The fire would char through these branches when it traversed the plains. Because the branches were green, only the bases would burn through. The alpha male found that these branches made perfect pointed digging sticks. Because they were green, they were more flexible than dry wood, and yet at the base the fire had hardened the wood to such a degree that it held a point very well.

The group liked fire. Of course, it had not always been so. Animals flee fire for the very good reason that fire doesn't care. It will burn anything that gets close to it regardless of size or strength. Over the millennia, those that ran from fire survived. Those that didn't, didn't.

Our *Homo erectus* ancestors at one time also feared fire. Caught on the open grasslands, fire would sometimes overtake the entire band, and even climbing trees didn't always allow for survival. If the flames were hot enough and high enough, the climbers would perish.

But this troop actually looked forward to fire. They knew that the beach didn't burn, and when smoke appeared on the horizon, everyone made a beeline for the shore and to the safety of the water.

It was after the fire had passed that the real fun began. There were insects singed by the fire or even cooked and clearly visible in the ashes. There were lizards, rodents, all types of goodies, and, best of all, no predators. Always with the fire they ran, always. They ran, even from their own kills, which made scavenging that much easier.

It was the alpha male that made the first connection. In the beginning, it was difficult because the red dancing flowers could hurt quite badly if one wasn't careful. But fire was fascinating also. Was it alive? It gave the appearance of eating wood but not like normal eating. It was warm but only up close. It was now red, now black, and then white. It grew and died quickly. It only affected certain things. Dry wood it ate, green it sometimes ate and sometimes not, and stone it ate not at all. It was truly fascinating.

The male liked to play with fire. At first, he would only poke at it with his digging stick, turning over the burning wood, prodding the ash to see where the fire was. He would wave a smoldering brand in the air to make the red flames jump out of the black char. He would place small twigs on top of burning branches to see the red devour the wood. The big male found fire to be the most fascinating thing in his life.

The problem was that the fire kept going away. A wall of flame would traverse the plain, black smoke rolling and animals fleeing, then it would be gone. The plain would be black and quiet. Here and there, a small plume of white smoke would emerge from a stump or a downed tree, but the vast wall of red would be gone or clear over on the horizon and way beyond reach. The male could make the stumps erupt into red but only as long as he fed the fire with wood. Then one day, he had an idea, and everything changed.

The shoreline occupied by the troop as well as the small sandspit used for sleeping was littered with driftwood. A river six miles to the south dumped lots of downed trees as well as branches and limbs into the ocean. The currents brought all this detritus north and, sweeping in shore, dumped a lot of it on the beach and sandspit. The male would use this wood to keep the fire alive. One evening after a fire on the plain, he brought a burning branch back to the shore and thrust it into a pile of driftwood. Almost immediately, the dry tinder caught fire, and soon the whole troop had to back away from the blaze because of the heat. The troop watched for a while then made their way out to the spit to sleep.

In the morning, the male came ashore and looked at the ashes left from the driftwood fire. Here and there wisps of smoke drifted up from the burned wood. The male turned over a log and red sparks snapped and popped. Quickly he grabbed a handful of twigs and put them on the log. A small tendril of smoke drifted up, then more, a twig snapped and charred. In a minute, the pile burst into flame. More twigs, then branches, and soon a roaring fire was burning.

The chill of the water and the coolness of the morning was offset by the heat of the fire. It was nice to wake up, wade across to the shore, and make up a nice warm fire to start the day. And that is exactly what the troop did. Each evening, they would use the ashes of yesterday's fire to start another. Over time, they learned just how much wood to ignite, how to arrange the logs, and what burned well and what not so well. One day it rained, and they lost their fire. But they remembered. And when the next lightning started, fire swept the plains; they took part of it back to the beach and again looked forward to warm mornings.

One evening, as the group was preparing the evening fire, a pack of hyenas showed up. Normally the whole bunch would have dashed into the water, but the hyenas seemed reluctant to attack. One of the males picked up a flaming branch and threw it at the pack. Sparks flew, and the hyenas scattered. Another brand, and the animals fled. It wasn't just fire on the plains that predators feared. It was any fire.

Homo erectus learned. They learned to keep fire—first brands, then coals in large clamshells, shells filled with punky wood and two or three coals. Fires that would keep when the two halves of the clamshell were bound together. Fires that had to be protected from water, not just rain but anything wet. Even a heavy damp would permeate the punk and snuff out the troop's survival mechanism.

There is no doubt that the subjugation of fire put enormous strain on our intellectual capacities. Those groups that could do a better job of holding on to fire prospered. Those that lost fire or were never able to master it were selected out. Brainpower prevailed. *Homo erectus* prevailed and became the first primate to radiate out of Africa. Our ancestors had arrived.

I'm sorry if this appears as a compendium of "just so" stories. The events recorded may or may not have occurred. However, at one time in our past, we looked much like modern-day chimps. Over a period of time, we became bipedal, naked, long-haired Homos. Did it happen on the plains of Africa, or in the ocean? No one knows for sure. But at the present time, the aquatic ape theory appears to be much more promising in explaining the present human condition than any other theory.

Certainly at one time in our past, our ancestors used unrefined tools, just as chimps do today. Then at some point, one individual (or more) discovered that by knocking off flakes, a better tool could be made, life would be improved, and we became not only tool users, but also toolmakers.

Again, at one time, our ancestors did not use fire. Then one (or more) groups discovered the value of the red flowers. Our future was then shaped by how well we were able to use and control fire. It is ironic that our discovery and use of fire occurred only because of ancestral pyromaniacs. Today, pyromania is punished. It is, however, possible that at one time, our very survival may have depended on pyromaniacs. Our survival also depended on our ability to make tools (it still does today). It was only because we had the brainpower to do these things that we survived.

But it was not so that we could survive using tools and fire that our brains changed. Yes, once the power of the brain came into evolutionary existence, we could use it, but our brain evolved the way it did because of bipedalism.

The theory of intelligence is so outrageous, so unbelievable, that it is understandable that many think of the whole concept as the result of some sort of special guidance. A group of apes (and at this time there were lots of apes) was forced to use the ocean as a refuge. They didn't want to, they wanted their trees back, but the choice was between extinction and ocean. They survived, and selection was for individuals that could wade into deeper water. Over time, we went from four legs to two arms and two legs. In the process, our pelvic structure rotated to accommodate our new stance. A chimp pelvis is long and extends up almost to the rib cage. If our pelvis maintained that shape in our aquatic ancestors, it would have stuck out the back like a porch. Selection was for shrinkage, and that's what happened. The shrinkage was not just on the outside. To maintain strength, the bones also shrank on the inside, making the birth canal opening smaller. Selection was for heads more round than oval, and we lost our muzzle. With the loss of the muzzle, our canine teeth became less valuable as weapons. In our sojourn in the sea, we became bipedal and defenseless. Our only weapon in the game of survival was our changed brain, which was a direct result of becoming bipedal. We are smart today because a grove of trees died in Africa around eight million years ago.

The whole process, the whole concept, is so peculiar as to defy reason. The method by which we became "we" is so unlikely, so bizarre, that we should not be surprised that a majority of the "we" do not believe it.

It was a series of accidents and luck—luck by the bucket; we could have very easily gone extinct any number of times. The lions could have gotten us. Sharks could have gotten us. Our changing brains might not have equipped us with the ability to go from the water back onto the land. Those of us who were the strongest might still be splashing around in the Indian Ocean.

We have a problem. Patterns in prehistory do not reflect our concept of reality. We are sure that intelligence is inevitable. We are convinced that brainpower is the culmination of evolution. We are positive that the human brain is Mother Nature's most cherished possession, at least that is what Richard Leakey claims. Trouble is the fossil record tells a different story.

We have *A. afarensis*, an early hominid some 3.6 million years old. It stuck around for seven hundred thousand years with virtually no change in brain size. Then we have *A. africanus*. It also existed for approximately seven hundred thousand years with no change in mental capacity. *A. aethiopicus*, half a million years. *A. boisei*, almost a million years. *A. robustus*, four hundred thousand years. All survived for hundreds of thousands of years with little or no change in the size of their brain.

Intelligence is supposed to be the be-all and end-all of evolution. We are supposed to be striving for bigger brains, seeking intelligence in the larger head—the meatier cranium. But our hominid ancestors did no such thing. There was no selection among these species for greater intelligence. If there had been, we have enough of a fossil record to validate such a contention.

And then there is *Homo erectus*, the original out-of-Africa man. Over time, this species of hominid spread from its ancestral home in Africa to Europe and Asia. They were the first truly cosmopolitan species. These hominids colonized most of the contiguous prehistoric world. And they did it without an appreciable increase in brain size. The first *H. erectus* fossil finds were discovered in sites 1.7 million years old. The species hung around, very successfully, for a million and a half years without any additional mental capacity. We can only hope that we will be as successful.

All these hominids lived, reproduced, and died on the plains and woodlands of Africa. They were very successful without showing any signs of selection for more brainpower. One even managed to expand even beyond the range of almost any other large mammal. And all this striving and reproducing was done with no indication at all of the need for greater intellectual capacity.

To be sure, the fossil record does show an increase in cranial capacity. *A. africanus* did not have the equipment at an average of

450 milliliters of brain size that *H. erectus* had at 940 milliliters. And *H. erectus* could not compete with modern humans at 1,450 milliliters of brain size. What we do not see is *A. africanus* getting bigger and bigger brains until they shade into *H. erectus*, which then shade into modern humans. Part of that last statement is not quite true, and the reason is interesting.

It is true that *A. africanus* exhibited little or no increase in brain size over the millennia. *H. erectus*, too, showed little or no increase until the very end, and then we see *H. erectus* starting to shade into archaic *Homo sapiens*.

And that is the problem. If brains are so important, so necessary to us, why did these primitive ancestral types exist without any selection for more brainpower? The answer is quite simple. We did not need additional brainpower to survive and reproduce at the time these species existed. *A. africanus* found a niche among the trees of Africa and, to a greater or lesser degree, prospered. I doubt *A. africanus* was ever a very common animal unlike pig or zebra or elephant. This is because we occupied a rather precarious place in nature. We needed the trees as a source of food and refuge, but the trees came and went at the whim of climate. And so *A. africanus* came and went at that same whim. If the land dried out, if the rains failed, then the trees died, and so did we. Better times, more fruit, and we survived, but not in huge numbers and not by much. And so for seven hundred thousand years, *A. africanus* existed, survived, and reproduced successfully with only 450 milliliters of brain. This is not much larger than a chimp's brain, and come to think of it, chimps do not show any tendency toward greater brainpower over millions of years either.

H. erectus lived and reproduced and died for well over a million years without any substantial increase in brain size. They didn't need it because *H. erectus* had fire. Fire made *H. erectus* a top predator. Every other animal fears fire, and it was fire that allowed *H. erectus* to leave Africa. *H. erectus* had an average brain size of 940 milliliters or about 1.8 pounds. This was enough to master fire and survive.

There was no need for more until *H. erectus* filled the habitable world. A half a million years ago, we began to see change. We began to see evolutionary pressure to enlarge our cranial capacity. As long as our *H. erectus* ancestors had places to go and things to do, they survived quite satisfactorily. But once all the available niches were filled up, things changed.

And this is why it is not exactly true that we do not see increased brain potential in *H. erectus.* For a long time, there was no change; a million years or more, then the selection for more brainpower began. We had run out of room to roam, and now we had to compete with each other. It is no coincidence that we started to change in Africa. Africa was the first place where we ran out of room. In competition with both prey and predators for a million years, we did fine with a 1.8-pound brain. In competition with each other, smarter meant better survival potential, and so smarter was selected for. We were our own worst enemies, and our big brains are a direct result of that enmity.

If, initially, there was no selective pressure on *H. erectus* and no pressure on *A. africanus* to enlarge their brains, then how did we get here? Where did the larger brain come from if not from one of our primitive ancestors? If one did not evolve into the other, if they did not evolve at all, then how did they get onto the plains of Africa in the first place? To find out, we must go back to the sea.

When we last left our aquatic apes, they were splashing around in the shallows on the east coast of Africa. As I have said, at the sea-land interface, we were fairly successful creatures, and soon we occupied the entire habitable coastline. But unlike the unlimited land area to the west and the unlimited sea to the east, the sea-land interface was limited. The available niche for our use was restricted to a narrow ribbon of habitation.

Where we could, we prospered and grew and multiplied and were successful. But soon a problem arose, a problem in paradise— overcrowding. We outgrew our habitat. We came from the trees, we could go back to the trees, and there really was no alternative. The beach was getting crowded.

I am sure it happened many times in our past. A group of aquatic apes would be forced from their seaside sanctuary by other stronger groups and forced to forage inland. Eventually, some groups abandoned the sea altogether, foraging and living in the trees away from the beach. Most of these attempts were unsuccessful, but a few such as the *A. afarensis*, *A. africanus*, and some others managed a successful transition to the veld. They had sufficient brainpower to survive, enough smarts to reproduce, and, once in the woodlands, plenty of space to expand, so there was little or no evolutionary pressure to increase brain size.

Meanwhile back in the water, the pressure was on. The hammer of occupying limited space beat upon the anvil of natural selection, and because we had lost our natural weapons, the only possible selection was for more brainpower.

And so another clan forced from the ocean sanctuary moved inland with slightly more mental capacity. This continued until the arrival of *H. erectus*. With *H. erectus*'s brainpower and fire came a populating of the known Earth. It is entirely possible, even probable, that *H. erectus* is the species responsible for the demise of the aquatic ape. After all, *H. erectus* would have had little or no fear of water.

We do not know at this time the relationship of our hominid progenitors. They exist suspended in time because it is impossible to tie them together. The reason they cannot be connected is because they are indeed separate. Each wave of losers invaded the land, survived, died, and left a few fossils as a reminder of their presence. These are all we have left of our hominid ancestors.

And now back to our *Homo erectus* forebearers. *H. erectus*, as I said, existed for well over a million years with little or no increase in cranial capacity. But when all the available habitat was filled, new population pressure was brought to bear. When *H. erectus* could solve overcrowding by expanding his domain, he did so. When there were no more uninhabited habitats, *H. erectus* faced a new selective force. Individuals better able to survive in competition with other *H. erectus* were selected for, and we see *H. erectus* responding just like the aquatic apes of yore. Larger brains, more smarts, and better survival potential breeds larger brains. It was all we had.

Our present mental capacity was not predestined. It was not the be-all and end-all of evolution. It is entirely possible that had this planet consisted of innumerable small islands more or less evenly distributed around the face of the globe, we might today be aquatic apes splashing around in the shallows, eating "cockles and mussels, alive alive-o."

CHAPTER 8

In 1996, a skeleton was discovered in the shallows of the Columbia River at Kennewick, Washington. The skeleton was remarkably complete and, at first, was thought to be that of a recent burial. Carbon 14 dating eliminated that idea when it showed that the bones were 8,400 years old. And the bones told a story.

Kennewick Man, for that is how he came to be known, was about forty-five years old at the time of his death and stood about five feet, eight inches tall. The bones had a fascinating tale to tell. The man had an injury to his left elbow. The wound could have been made by a right-handed individual. A blow to his chest had been delivered with such force that the rib bones had broken and were separated from the breastbone. He had a skull fracture, again on the left side of his head. All these wounds showed some signs of healing. Kennewick Man was a tough old codger and had lived a violent and traumatic life.

But the most egregious injury to Kennewick Man was in his pelvis. A spear point was embedded in the pelvic bone, and it, too, had healed over. This meant that it had been there quite a while. The skeleton of Kennewick Man vividly demonstrates the violent lives we lived in what we would like to think of as our peaceful past.

Then there is Otzi the Iceman. He, too, is old. Otzi, like Kennewick, was an adult male, and his body, too, told a fascinating story. Discovered in 1991, Otzi was not just a skeleton but a mummy. He was preserved in the ice of the Alps for more than five thousand years. It was not until 2001 that the cause of Otzi's death was discovered. It turns out that

Otzi had an arrowpoint embedded in his left shoulder and a debilitating wound to one hand. Otzi suffered a violent death.

Now what are the odds that the two most complete remains of our ancestral progenitors ever to be found are the only ones that just happen to have projectile points in them? Is it possible that Otzi was cleaning his bow when it went off accidentally? Is it possible that Kennewick Man was running with his lance in his hand when he tripped and hurt himself? What are the odds that Kennewick and Otzi were the only individuals that bore indisputable evidence of violence in their particular population mix? What are the odds that these two sets of remains of early man are the only two that just happened to contain projectile points and are the only two that we discover? The odds are against it. It is far more likely that our ancestors lived violent, brutal lives. It is far more likely that an individual would die a violent death rather than from old age.

We want to think that our species is naturally nice. We see our ancestors as nice guys. We have this vision of the peaceful native, the noble savage, living in harmony with his environment. He takes only what he needs. He lives a quiet life. He leaves his habitat as he found it. Our ancestors were, according to academics, the perfect species.

It is delusion. We lived lives that were short, brutal, and dangerous. Those that were ready and willing to do whatever was necessary survived. This is reality. This is how we lived. And for many, this lifestyle still impacts us to this day.

Dawn on the plains of East Africa. It is the middle of the dry season, and green is turning brown, water holes are shrinking, and the sky is a clear golden blue lit by a sun, bright and hot. The plain is not, strictly speaking, a plain. True, it consists of rolling hills stretching to the horizon, now green, now brown, depending on water. It is, for the most part, treeless. But there are, sprinkled like raisins in a pudding, rock-covered mounds. Black stone, giant boulders piled chockablock, full of holes and caves, and, oddly enough, trees—grotesque twisted trees with naked roots clutching at the stone, dipping into holes and cracks and desperately trying to survive.

The trees cling to the rocks because the rocks are where the water is. During the wet season, the water drains off the plain, but on the rock-covered mounds, It seeps Into the cracks and crevices, and there it stays. It stays until tree roots tap this wet treasure.

Cheetahs make the rocky mounds home, leopards, too, sometimes. Hyrax hide in nooks and crannies. The occasional cobra and meerkat come together in their herky-jerky minuet among the rocks while the ever-present hawk uses the tallest tree on the mound for a lookout and, perhaps, a nest.

But not this mound or kopje, as it is called. The hyrax, the hawk, the cheetah, the leopard—all missing; only the occasional cobra remains among the tree roots. The reason lies on the north side of the kopje, where a single huge boulder, black and rain-stained gray, leans as if to break away from the rest of the mound. The crag is forty feet high and some twenty-five feet wide, slanting to the northeast at an angle of about sixty degrees to the horizontal. Below the overhang is a small hollow, the width of the rock and maybe twenty feet deep. Here in this shallow depression lay the reason the mound is abandoned. This place is devoid of wildlife because it is the home of man and fire.

These are fully modern humans, not Neanderthal, not Archaic, not primitive, but fully modern in body size and brain shape. The group consists of 9 adult males, 14 adult females, and 25 children under the age of 13. The males are 5 feet 6 to 5 feet 8 and weigh 130 to 150 pounds. The females average 5 feet, 2 inches and weigh a little over 90 pounds. All are black, all have head hair in what we would today call dreadlocks, and the males sport beards, which on older individuals were so large and bushy that from a distance the head is almost as wide as the shoulders.

Most of the individuals are naked except for a small flap of animal hide suspended from a cord tied around the waist. The flap of skin is not very big, just large enough to cover the genitals, which is its purpose. There is no modesty here; the covering is not big enough nor well enough secured to promote privacy. Its purpose is entirely practical.

The fire is hot at times when too much wood is used or when a particular kind of wood is used, a kind that burns very hot because of the high level of oils present. This small flap of skin protects the delicate genital area from the heat when one squats in front of the fire or when hot coals pop, snap and fly off that same fire. Our first garments were worn for practical purposes rather than modesty.

At first light, the group is up and active. On the side that clusters the females someone stirs the fire while several step out into the open, maybe forty or fifty feet from camp, to defecate and urinate. Some of the men do the same while the children stay closer to the camp.

The sun is just up, and the advantage of the campsite quickly becomes apparent. The sun—weak and watery at this time of day, an angry red through distant gauzy clouds—slowly rises. All take in the first faint warmth of the sun, basking in its heat and allowing the light to stir cold-soaked bones. Someone stretches and burps. Several late risers slowly meander out of the camp to a point some sixty feet away. While two relieve themselves, others look around, scanning both the mound and the sea of grass in front of it. There is not much of a chance of danger, but it is best to be on guard. A pride of lions may just be returning from a night hunt, or there may be a pack of hyenas heading home. Sixty feet is an awfully long distance, and the fire is the only sanctuary.

The sunlight is now warm, and the chill rapidly dissipates. Life at Black Rock camp really starts to stir. First, the fire must be tended to. There are three fires in a rough semicircle around the outer perimeter of the overhang. The rain, when it comes, always makes its way out of the south, so the fire can be close to the outer edge of the shelter cast by the boulder and still be sheltered from the rain. The fires are fed wood from the many trees that, over the years, have lost the battle with the rocks and died. Soon, a number of individuals are feeding the flames.

And there is a pattern to the feeding. On one side is a fire tended by females and young children, both male and female. On the other side there is a larger fire tended by the adult males. In the middle is the smallest of the fires, tended by five boys; the youngest is nine, and the oldest is twelve. Each group tends its own fire. Each group stirs and arranges and feeds its own fire. There is one exception: an old crone of at least forty years and crippled from a lifetime of sleeping on cold, wet ground, from wounds old and some not so, and from broken bones. She is responsible for keeping all the fires alive at night and when the clan is foraging. Because of her status as fire tender, she has the right to any fire. The old woman has yet another talent used by the clan—knowledge. She has been there and done that and can be relied upon for sage advice, taken or not, about many things. This, plus the fire keeping, gets her entry into any fire and any food prepared there.

The women and children forage every day. They now set out twelve females and eighteen children. Each female has a short digging stick, two spears about five feet long, and a chunk of rock about the size of a woman's palm shaped like a teardrop. The digging stick and spears are carried in one hand while the stone rests in a small hide bag tied to the cord around the woman's waist. Four of the women have something else hanging from their waistband—shells of a freshwater clam about six inches long. Each woman has a pair of these shells bound together with strips of hide. From the bound clams comes little streams of smoke, and the odor of burning wood hangs heavy in the morning air.

The group moves out to a smaller kopje about a half mile from camp. The women and children move in one cohesive group with adults on the outside and children in the middle. All are on the lookout not just for predators but for snakes which, if found, will be killed and eaten. The march is purposeful and direct, and in less than fifteen minutes, the group has reached the base of the small pile of rocks.

The women work in pairs or in threesomes. Two women start to build a fire, and the children carefully search out and bring wood to a pile next to the fire starters. One of the women takes the clamshell from her belt and unties the laces. Carefully she opens the shell to reveal a lump of glowing charcoal nested in a bed of dry punky wood. Gently she blows on the coal, and it snaps to life, giving off little red sparks. She tips the coal out, carefully preserving the punk. The coal lands in a bed of dry grass, and two or three gentle puffs of air bring bright red flames to the tinder. Twigs are quickly added, then small broken bits of wood and then larger branches, and in less than a minute, the group has a good-sized fire going. In an emergency, a fire starter can go from glowing coal to fire in twenty seconds. This is a skill of no small import when a pride of lions or a pack of hyenas is spotted.

The group begins to forage in the vicinity of the fire. The fire is the key. Without it, the stone hand axes and the pointy sticks would be almost useless. With fire, these are not so much weapons as tools, digging sticks, and bark scrapers. Bulbs are found and dug up when their dry tops give them away. Insects are caught and crunched by the children, while termite fishing takes up a good deal of the morning.

The children attack the termite mound with their own pointy sticks, finding weak spots in the structure and forcing a hole into the mound. Then a thin grass stem is inserted into the hole. Termites inside bite the stem just as they would anything invading the nest. The stem is then pulled out, quickly inserted in the mouth, and the insects are crushed before they can bite a lip or a tongue.

The band will spend the next three hours foraging. In that length of time, most of the group will get enough food to take the edge off their hunger and supply the bulk of their daily caloric needs. Most of this consists of roots, tubers, bulbs, fruit, and buds with a small amount of protein in the form of insects and small animals. Too quickly, the sun is up high, and it is getting hot. The group is about ready to head back for camp. The alpha female and her two partners lead the way.

It is this relationship of female to female that is crucial. All the females have established relationships with one or two other females. The pair or threesome do everything together. They sleep together, they eat together, they work together, and they even defecate together. Early on, bonds form among the females. Girls as young as four or five develop relationships with one or two of their peers. These bonds will remain strong for the rest of the individuals' lives. Such relationships are critical to a species that has little, save its own wits, to count on for survival. From an early age, they watch each other's back, help each other forage, assist in birthing and child-rearing and in maintaining social status within the group.

A trio of females leads the female half of the clan. This is unusual in that the one older woman at twenty-eight bonded to a pair of females aged seventeen and nineteen. The three are always together, especially when one of them is in the last stages of pregnancy or in the first weeks after childbirth. Our trio is inseparable.

Inseparability is the key to survival, and status. Three pairs of eyes watching means danger is much more easily spotted. Because of the older female's experience and knowledge, she is looked up to by most of her campmates. As a result, the trio dominates the female side of the group.

The bonds are deep and lasting. The attachment one female feels for her companion or companions is substantial. Each knows what the other can do and how they will behave. The bond is strong but unspoken. Bond mates are satisfied to be in the presence of the others, content to feel the delicate fingers of the others searching for lice among the dreadlocks. They find pleasure in resting comfortably with each other. It is enough to know of the others' presence, maybe a pat or two of encouragement and a grin of acknowledgment.

It is past the middle of the day now. The sun is well beyond straight up and is getting very hot. The dominant female glances around, and upon some unspoken command, the whole group begins to move back toward base camp. The distance is covered at a rapid clip, and the children must run to keep up. It is at this time, between fires, that the troop is most vulnerable. At camp, the females rest before beginning the unending round of chores necessary to survive in a hostile world.

The men, too, set out shortly after sunrise. But their goal was different. They are after Cape buffalo, one of the most dangerous animals on the plains of Africa.

The men hunt and the women gather. Sexist? Not really. The females have to take care of the children. The children take a long time to grow up, and females are saddled with this responsibility for most of their childbearing years. No sooner is one—at four or five, able to forage on his own (with supervision, of course)—than another child is at the breast, helpless and in need of constant attention. Hunting game, large or small, is almost impossible under such a burden. As we shall see, female hunting is really not all that necessary because our forefathers were the top predators on the plains of Africa, even killing elephant on occasion. No mean feat for someone with nothing more than pointy sticks and shaped rocks.

The men did not hunt buffalo in the classic sense of chasing and spearing. The buffalos were too wily to let our hunters get close enough to do any damage. No, what was needed was guile, intelligence, and a water hole. The tribe didn't go to the water hole for water. They didn't need to. The same rocks that collected and stored water for the trees also produced potholes high up in the mound, potholes that always contained water. This was where the tribe drank, not at the dangerous and somewhat distant water hole.

The men set off at a trot, the standard speed for purposeful travel, and in less than thirty minutes, they arrive at the site. The water hole is about eighty feet wide and two hundred feet long. It is shaped like a lopsided hourglass lying on its side with two deep pools, one to the east and one to the west, connected by a shallow waist. Around the pool are clumps of small trees and brush separated by the paths the animals use to come down to the water's edge. The water is a dark cloudy brown, and the shore, for a distance of ten feet back, a black, smelly, muddy muck. Beyond that is twenty to thirty feet of bare ground trampled daily by thousands of hooves. Then there are the clumps of trees and brush, perfect for ambush by any predator. And that is exactly what our little group of nine plan, an ambush. But not just any ambush, no, a special man-made ambush.

The buffalo graze from first light to hot. When the sun gets up in the sky, the animals come down to the water hole for a drink. Afterward, they will lie in the shade if they can find any and chew the cud. Late in the afternoon, another round of grazing, and then a night of cud chewing and fitful sleep.

The hunters have about three hours to set their trap and will need all of it. On the north side of the water hole are two clumps of trees about thirty feet apart, separated by one of the pathways leading down to the water. Here the men will set their ambush.

The task is simple: dig a trench one hand-width wide and one arm-length deep in the soil between the two clumps of trees. Digging isn't easy in the hard soil packed by decades of hooves, but the pointed sticks and the sharp teardrop-shaped stones help. It takes two hours, but in the end, the men have constructed a straight-sided trench almost exactly six inches wide and more than eighteen inches deep from one clump of trees to the other. The excavated soil is scattered in the brush except for a small pile kept on a piece of hide.

Next, the men gather twigs from the brush and lay them across the trench. Then comes leaves, and soon the trench is covered, obvious but covered. The saved dirt is then gently placed on top of the leaves, and after they are finished, it takes a very sharp eye indeed to notice the trap. The men now urinate on the ground between the trap and the water, some going as far as the muddy black muck to piss. Soon the whole area reeks of urine, masking the odor of man. The trap is now set.

The buffalo come down to the water hole out of the northwest. The breeze is out of the south-southeast, and most animals approach the water hole from downwind. This could cause a problem for the trap if some animals came at it from the wrong way at the wrong time. So two of the men station themselves in the brush just north of the clump of trees where the trap is. Animals coming down to drink will see and smell the men and circle around them.

Just west of the trap is another broader path to the water hole, and it is this avenue that the buffalo will use.

The men station themselves to the east and west of the water hole far enough into the trees to be unseen and un-smelled. This is the dangerous time. Fire cannot be used, but there are trees handy to climb, and some of the men do so, not so much to escape predation but to better see the approaching animals.

Almost immediately, a problem arises. From the south come the sounds of elephant. A small herd of about twenty-five animals led by an old female is on its way down to the water hole. If the animals catch wind of the hunters, it is likely that all the men will have to scatter, and the small trees by the water hole will be of no benefit. But the elephants are concerned with water only, and in fifteen to twenty minutes, they have slaked their thirst and moved off in the direction from which they came. The elephant is the one unknown in tribal life. Had the animals come from the north, they would have discovered the trap, caught the man scent, gone on a rampage, and destroyed the morning's work.

The sun beats down on the water hole, cicadas whistle in the shrubs, birds fly in individually and in flocks to drink at the water's edge. A small herd of impala nervously approaches, catches a stray man smell, and immediately scatter. They need to drink but not now. A family of pigs trots down to the shore but leaves in a hurry after one of them spots something moving in the brush. One of the men had shifted his position under the shade of a small tree. After that, nothing happened for almost an hour. One of the reasons that our hunters set their traps at this time of day was that the buffalo were one of the few animals to water at midday. Most traffic was early to midmorning or late afternoon when the heavy heat of the day lifted. Buffalos didn't care. They watered at midday then retired to digest and rest.

Finally, just when it seemed the animals had picked another water hole, they came. A herd of about fifty—mostly cows and year-lings, snorting and suspicious—comes out of the northwest exactly as the hunters had anticipated. The older cows stop at the beginning of the pathway to the water. Equal amounts of suspicion and caution dictate their actions.

The lead animal is an old cow. She stands transfixed, glowering at the scene. Something isn't right. Something is definitely wrong. Something. But what? The old cow can't make up her mind, and behind her the herd is getting antsy. They can smell water and are thirsty, having had nothing to drink since yesterday at this same time. The animals bunch up in the pathway behind the old cow, swinging their heads back and forth, snorting and pawing, ears waving like semaphores, and anticipating water. A young cow takes a couple of steps forward, then another and another. Soon the whole herd is standing hock deep in the black ooze at the edge of the water hole. They move to the middle of the hourglass where the water is shallow, and they begin to drink. The men edge closer to the animals and wait patiently. The timing has to be perfect; the bulk of the animals need to have their fill of water. But if the men wait too long, the herd will spread out and scatter when the trap is sprung. If they act too soon, the herd will take off back the way it came, so timing is critical. Just as the animals start to mill, the leader of the hunters gives the signal, and nine screaming banshees converge on the herd from east and west.

The animals panic and head for the only path available—straight into the trap. Older animals have been through this before, but the screams and the waving hides make danger appear much bigger than it really is. The whole herd turns on a dime and stampedes up the path between the trees. Most of the animals make it, some actually jumping over the unseen danger. Some put a hind leg into the trench, stumble, pull the leg out, and thunder on, none the worse for wear. For a few seconds, it looks as if all the buffalo might escape. All the work would have been for nothing, and the trap destroyed.

Then a young cow on the edge of the herd puts her left front leg squarely on top of an undamaged length of trench. The cow bawls in terror as her leg, with all her weight behind it, travels the eighteen-inch distance to the bottom of the trench. Her own weight, coupled with the pressure of animals behind her, causes the cow to do a forward somersault, landing on her back half on the ground and half on the backs of the animals in front of her. With a very audible crack, her humerus snaps just above the elbow. Three or four animals run over her before she can scramble to her feet, bleeding from several cuts made by the hooves of her herd mates. Her left front leg dangles helplessly. Hobbling on three feet, she starts after the herd, only to be stopped by three specters screeching and waving animal hides in front of her face.

On the other side of the herd, a yearling bull does the same thing as the cow, stepping into the trench, falling forward, and breaking his right front leg. He, too, watches as the rest of the herd disappears in a cloud of dust onto the plain.

In less than twenty seconds, it is all over, and the hunters have enough meat on hand to keep them for a week. The animals will have to be killed, but that is easy. Three of the men stand in front of the cow dancing and prancing, shouting and waving their hides. The cow, panic-stricken, tries to keep the three men in front of her. It would have been easy on four legs. She might even have lunged at one or the other of them. But with only one front leg, she cannot turn except by a hobbled hop that takes time and lots of effort.

All of a sudden, she lets out a frantic bellow of pain. A fourth hunter, taking advantage of her distraction and her useless leg, has slipped up behind her and driven his spear into her body just behind the rib cage. Buffalo hide is thick and hard around the shoulders. But on the sides and belly, it is thinner, and fire-hardened spears can penetrate. The weapon is driven forward at least two feet, seeking the heart or, barring that, the lungs. Another spear on the other side causes the cow to jerk that way, and overbalanced on one front leg, she topples over. Before she can scramble to her feet, two more spears are thrust into her exposed belly. Her nose runs pink with frothy lung blood. A fifth spear finds the heart, and the beast dies.

The bull is handled in the same manner, taking eight spear thrusts before a fatal jab finally kills it.

Not all the men involve themselves with the killing of the buffalo. One, older and not so quick or strong, keeps a lookout on the northern side of the water hole while a companion watches from the south side. As the men finish off the bull, the man on the south side of the water hole gives a sharp yipe that all know means that hyenas are approaching. Two men break off from the killing and quickly start a fire using the same method that the women had used. In less than a minute, small flames start to crackle and rise. By placing the fire between the two dead beasts, the men are able to protect both while they butcher the animals.

One of the men grabs a burning branch and thrusts it into the dried grass and brush under the clump of trees that anchored one end of the trench trap. Quickly the flames spread, eating the small twigs and clumps of grass but not getting high enough to burn the trees. This burning will scare off any predators and eliminate a place from which potential attackers could come upon the men unobserved. The hyenas, at the first sign of fire, slink off, whimpering and crying their frustration.

Now is the best time, and the men take full advantage of their success. One prances and dances in front of the dead bull, poking at it with a spear and reciting his role in the killing. Another spreads his arms wide and screams his satisfaction to the heaven. All are on an adrenaline high, and all will take a few minutes to calm down. But this is what it is all about; the rush, the excitement, the danger—real or imagined—all contribute to this moment, this special time, and the men revel in the glory of the kill each in his own unique way.

And now they are hungry. It is well past noon, and none has eaten anything beyond some old rotten meat the day before. Quickly they set to work. The animals are rolled over onto their backs, a long cut is made down the gut, and the innards are dragged out. At the same time, the spears are withdrawn, examined for damage, and placed aside. Comments are made with respect to the ability of the spear owner, depending on how well or not so well the thrust had been made.

The men are after the liver and kidneys. The liver comes first. Lying in a pile of guts, it is quickly cut free and sliced into strips. The kidneys are in a caul of fat on either side of the backbone just behind the rib cage. These are quickly lifted out and passed around. The men are soon enjoying fresh hot raw liver and kidney. The liver is rich and flavorful, and the kidneys have a decidedly salty flavor. In a few minutes, two livers and four kidneys are gone.

Hunger temporarily slaked, the men set to work to complete the butchering of the two animals. Front legs first, and it was easy as the leg is attached to the body only by skin and flesh. Beginning at the point where the leg meets the breastbone, a few quick slashes with a sharpened flint axe and the leg comes off. Next the hind leg, harder but possible. They key is the bone with the round knob on the end. The men cut into the groin until this bone is exposed, then they follow it to the point where it enters the hip. By pulling and cutting, this bone is removed from its socket without having to be broken. Soon, only the barrel-shaped chest and the backbone are left.

But the hunters are not through. There is still more meat to come off. The cow is rolled over onto her belly—or what's left of it—and two parallel cuts are made on either side of the backbone from neck to rump. This is the hard part. The shoulder skin is the toughest, and the stone blades take a while getting through to the meat underneath. But the hunters keep at it, and soon two long strips of meat and skin the thickness of a man's thigh are added to the growing pile.

One last operation and the men obtain a small but choice cut of buffalo. The cow, now mostly a skeleton, is again rolled over on what is left of her back. Lying along the backbone just behind where the kidneys had been and running back to the pelvis is a small strip of meat about the thickness of a man's arm at one end, tapering down to nothing at the other. Quickly, two of these are ripped from each animal and laid aside. On the way home, the men will assuage their hunger on these tender strips.

Together the carcasses produce over one thousand pounds of usable meat and bone. Various pieces are bound with hide strips brought for that purpose and hoisted onto backs with a tumpline. The men could easily carry over two hundred pounds, and none of these loads were that heavy. With one last look of satisfaction, the men gather up their spears and abandon what was left of the buffaloes to the vultures and the jackals. The trip back is watchful but good-humored. The hunt has been successful. No one was hurt, and the supply of meat will last five or six days. But most importantly, behind the fire at camp, both they and their meat would be safe.

At camp, the meat is piled in a heap at the back of the overhang. Two of the back strips are hauled out, sliced up, and thrown on the fire. The men could eat the meat raw, but they prefer it cooked. Soon, the aroma of roasting meat fills the camp. Impatient at the cooking time, the men snatch half-cooked pieces of meat from the fire and scarf them down. In less than thirty minutes, the men put away between eight and ten pounds of meat each. This, on top of what they had already eaten at the kill site and on the way home, left the lot with bellies distended to the point of incipient pregnancy. Full, satisfied, and tired, the men do what comes naturally. They go to sleep. Curling up on old mangy skins, the nine are soon sound asleep and snoring.

The women go about their chores as the men dine, but do not attempt to obtain any of the meat for themselves. They know better. The men would have cuffed anyone attempting to steal their food. But now it was different. After giving the hunters a few minutes to fall asleep, one of the older females cautiously creeps over to the meat pile and removes one of the cow forelimbs. Gingerly, she makes her way back to the female fire. When no alarm was raised, another female steals another leg, this one a hind leg from the bull buffalo. And at 150 pounds, it was the biggest piece of meat on the pile. If you are going to take, you might as well take the biggest. If caught, the cuffing and kicking would be the same as it would be for taking a small piece of meat.

And now the women and children feast. This is where the occupants of the female fire get a good part of their protein. Soon, they, too, are full of buffalo, and most follow the males off into that dream world of full bellies and safety.

And thus it was. The group survived because the males were top predators, not because they were stronger or better equipped by nature to occupy the top spot, but because of their intelligence. We had at last come to that point in the survival game where our lone survival mechanism was paying off.

Once in a while, the boys at the middle fire took from the male meat pile, but mostly they hunted for themselves. They did not hunt buffalo but rather smaller game. All the boys had short spears and clubs. They also carried small palm-size pebbles that they could launch at springhares and small birds. They survived on ostrich eggs, lizards, insects, and just about anything else edible including some of the roots and bulbs the women ate and the boys themselves had harvested when they were at the female fire.

For the boys, life is tough. They are physically evicted from the female fire at eight years or so of age, at about the time they start to show some interest in the females. They are, however, too young to join the males. They will not be able to move to the male fire for five to six years. They must fend for themselves and can expect help only from their fire mates.

Life is not easy for the clan. They have only the glimmerings of intelligence with which to survive. That and fire. Fire is really the key. It is the first and most important weapon for both protection and warmth. Fire is the reason for our success. But there is something else, something that, unlike fire, our ancestors shared with many other African animals.

Many animals form relationships with others of the same species. Sometimes the relationship is a true pair-bond with the couple acquiring and defending a territory. Gibbons fall into this category. More often, the relationship is with the same sex. Male chimps sometimes form close relationships with other males. But the true benefit to this kind of bond occurs among male lions.

Two lions are better than one, always. Male lions are driven from the pride when they begin to display a mane. They must live the life of vagabonds until they reach their prime in three or four years. Then they will seek out a pride and challenge the resident male for access to the females. If the resident male is strong, powerful, and experienced, the challenger will likely lose. But if the male is older and not as quick as he used to be, the challenger just might win and take over the pride.

Sometimes two male cubs will be ejected from the pride at the same time. These cubs may form a bond, which continues through their three or four years of isolation. When the pair reach maturity, they will almost certainly prevail over a single male for pride control even if the resident male is at his peak of fitness.

Same-sex bonding is common in predators, and we were predators. The predisposition to form bonds runs deep in the human animal for the very good reason that such behavioral patterns had survival value. This was true especially among the females. As we have seen, we did not pair-bond with the opposite sex. We have never been pair-bonding animals. We were not, and are not, monogamous animals.

Our predisposition was to form relationships with individuals of the same sex. This kind of predisposition had tremendous survival value to our ancestors. If danger threatened, the bonded female pair would stand back to back, protecting each other. During childbirth and the first few days afterward, it was absolutely essential that the female have assistance and help. Childbirth 150,000 years ago was just as painful and debilitating as it is today. A long, difficult birth would leave the mother exhausted and at extreme risk. This vulnerability would have been offset to a degree by the assistance from another female who had strong feelings of attachment to the new mother.

Males, too, benefit from such relationships. Two males, in their prime, together could dominate the group by presenting a united front to any other male trying to achieve dominance. But the greatest advantage to such commitments is survival. Two people who look out for each other, who work together, sleep together (in a non-sexual context), and eat together substantially improve their chances of survival. There are two pairs of eyes looking for danger. There are two pairs of arms to fend off danger, and if one is hurt, there is someone to look after her, to assist in her recovery.

Among the males, such relationships would be expressed in terms of companionship. A bonded male would experience a sense of satisfaction and pleasure in the presence of his bond mate. The two would be comfortable together, at ease. A feeling of contentment and happiness would mark the relationship. A real sense of sorrow was felt if one or the other fell ill, was hurt, or died. Males did not bond with females; females did not bond with males. Our ancestors had much stronger feelings with respect to others of the same sex rather than toward others of the opposite sex. As we shall see, this predisposition regarding same-sex bonding has done much to muddy the waters of human behavior.

In the meantime, we shall return to our Black Rock group as another day dawns. Always it is the same: feed the fire, urinate, defecate, bask in the weak morning sun, cut a chunk from the by-now-rotting buffalo, climb back into the pile of rocks for a drink of water, then contemplate the day. This day, however, would be different. Alpha had decided to visit the boundary between the clan's territory and the territory of the neighboring Lakeside group. Signs, expertly read, told the alpha that four male members of the Lakeside clan had visited the same water hole that the Black Rock clan had used to ambush the buffalo, and the sign was fresh. The water hole was less than one half mile from the boundary, but it was clearly in Black Rock territory, and something had to be done. The alpha's partner, beta, would go along, as well as two additional pairs of males. An older male that had recently lost his partner and two of the males that had only recently joined the male fire and had not as yet acquired the skills necessary to invade enemy territory would remain in camp to guard the site.

With little preparation and even less talk, the six men started out for the boundary. The preparation consisted of checking the edges of the same stones used to dig the ambush trench and touching them up. They also inspected the fire-hardened spear points and grabbed a handful of rotten meat.

Soon they were six dots moving single file across a distant hill, over the top, and out of sight. No one in camp paid them any mind. The three remaining males would stay in camp and do the chores. The females, as always, took most of the younger children and began to forage, but closer to camp than normal. The five boys, tired of their own rotting meat, went out on a springhare hunt.

The six Black Rock males moved at a steady clip and soon were in Lakeside territory. The Lakeside clan had recently suffered two almost simultaneous disasters. Twenty days ago, the eight males of this clan had set a similar kind of trap for buffalo at a water hole a few miles from their camp. The men were in position and waiting when a male elephant in musth found their scent. The rogue elephant immediately charged the men, scattering them to the four winds. The only possibility of escape was to run in different directions and hope that the elephant would follow someone else. The men ran as pairs, companion sticking close to companion, and the elephant chose two of them to attack. The beast was much faster than the runners and quickly overtook them. With one swipe of his trunk, he knocked both men senseless and into the bush. The bull then trampled one while picking the other one up by one leg and slamming him into the ground, breaking every bone in his body.

Without hesitation, the bull whirled around and started after a second pair headed for a small pile of rocks where they hoped they would find a crevice deep enough to hide in. It was not to be. The beast caught up with them a good twenty yards short of sanctuary and killed each in turn.

The remaining four men were, by now, far enough away to be out of danger, but in less than three minutes, the Lakeside clan had lost half its adult male population.

The disaster suffered by the men was bad enough, but what happened a few days later was even more catastrophic. The dominant females in the Lakeside clan were a twenty-four-year-old and her twenty-two-year-old companion. Together they had guided the fortunes of the females for almost six years. The female alpha was both smart and strong, a natural leader and sure of herself. This surety had been reinforced by years of decisions that had resulted in prosperous times for the Lakeside females. Altogether, there were twenty-three females of breeding age in the clan. A very high number and a testament to the leadership skills of the alpha. The females, along with some of the older children, had even taken up hunting on a limited basis. This was a necessity because the eight males could not provide enough food for themselves and for all the females.

Today, however, all that would go for naught. The men made the mistake of allowing a rogue bull elephant to sneak up on them. The females' fate was sealed in the black thunderclouds that towered over the little park in which the women and children foraged.

Rain comes to the plain on a regular basis in season, and for the most part, the clan ignores the wet. Adaptation had to be made to rain, but only at camp. Rain is death to fire, and so for hundreds of years, the clan had built a sort of shelter over the fire to keep the rain out. Unlike Black Rock, the Lakeside clan had no caves or overhangs in its territory. As a result, one of the first chores when camp was set up was to construct a covering over the firepit out of logs, brush, and hides. Only in the last couple of hundred years had the clan started enlarging the shelter to a point where both the fire and the clan had some refuge from the wind and rain, and thus, housing was born.

Out on the veld, however, it was a different matter. A protection fire on the plains would go out in a heavy rain. The clan's response to this was to pile as much brush and wood on the fire as possible, hoping that even with a heavy rain, some of the fire would be protected. Besides that, predators did not hunt in the rain, nor did they do so immediately after. Lions don't like to get their feet wet, even when they already are.

The females did not normally seek shelter when the rains came. Rain, like fleas and lice, was something to put up with, something to endure. It even offered a bit of an inducement to bulb and root digging by softening the soil. It also sent most small animals and even insects to shelter. So rain was a mixed blessing, something the clan just had to live with.

But this rain was different. In all her years as leader of the troop, the alpha female had never seen a storm this large. She summoned the oldest female in the group and asked her about the approaching rain clouds. The old woman, at thirty-five, had a great deal of knowledge and experience, but the stresses of life had taken their toll, and she could not remember as well as she used to. Yes, she had seen a storm this size before, many years ago. Yes, it produced much rain, a lot of light flashes and loud noises just as this one was doing. It was unusual but not unique, and there was something else, something—but what? The old one just couldn't remember.

Reassured, the alpha female set about digging up bulbs as the first drops of rain came. But these were no ordinary drops; they were huge and smarted when they hit bare skin. The alpha immediately signaled the children to throw all the collected brush and logs on the fire. As the fire roared to life, the rain started in earnest, pelting down on the group. Gusts of wind whipped the blazing fire around, blowing red-hot streamers into the sky.

At this rate, the fire would soon go out, and the alpha knew it. Uneasy over the approaching storm, she gave the signal, and the group quickly gathered up their tools and weapons and started back to camp. They would go hungry this day because they had harvested little in the way of food, and the males were not sharing their meager stocks at all.

The group was almost halfway back to camp when the skies really opened up and sheets of rain began to fall. The wind whipped the droplets around until many were traveling almost horizontally. It was hard to keep the group together, and soon the women and children were strung out in a line while methodically plodding along.

Then it happened. A few white stones fell out of the sky, bounced along the ground, and came to rest on the wet red soil. The troop paid no attention. Camp was close, less than half a mile; soon they would be out of the weather. The hard white balls were larger now, some thumb sized, and when some of them landed on a bare back or arm, they hurt. The alpha sought out the old one and asked a question with her eyes. "Yes," the old one said; she had seen this once before. Like the last time, they were white, hard, and they were cold. But after a while they went away. Yes, as big as these and bigger; some of the clan were seriously injured the last time, and some even died. There was something else, something she meant to remember, needed to remember, but couldn't. It just wouldn't come.

The camp was too far, and the hailstones were creating huge splashes in the pools of water. The big ones really hurt. Two or three of the children were already bleeding from wounds caused by the hard white rocks. Something had to be done.

The women and children were only a little way from a small grove of acacias. The alpha female decided they would wait out the pounding in the lee of the trees. In a few minutes, the whole group was under the trees. Acacias are not really heavy foliage trees, but if the women stood on the side of the tree away from the wind, most of the hard white balls missed them.

The alpha and her partner, as were their due, took refuge behind the biggest tree and so were protected from the hail and much of the rain. Two or three other females of high rank joined her. The alpha squatted at the roots of the tree and cast a reassuring glance at her companion. No one had ever seen anything like this; it would be something to describe to the others around the night fire. The hail crashed through the branches of the trees, raining leaves and twig down on the assemblage below. The white flashes cut ragged wounds across the blackened sky. The mighty sound of…of—the alpha could make no comparison; stampeding elephants were not even close. It was wild and terrible, but also awe-inspiring and somehow a grand stimulating show. The alpha female stood up, spread her arms wide, and screamed her defiance at the sky. It was good to be alive.

Just then she noticed the old one who had taken refuge under another tree hurry toward her, shouting and flinching every time a hailstone hit her. The almost continuous sound of thunder and the pounding of the hail drowned out her words, but the alpha could see the old one had something to tell her. Unfortunately, the recovered memory came too late for both the alpha and the old one.

The flash was blinding, and everyone under the big tree was dead before the thunder started to roll. The tree was split from top to bottom, literally blasted open by the lightning. Shards of splintered acacia flew everywhere, the stink of sulfur burned the air, and smoke from innumerable burning embers quickly lost the fight with the wind and the pelting rain.

At the base of the shattered acacia lay the battered and burned bodies of five of the Lakeside clan's females. With them were the remains of ten children, from babes at breast to a couple of nine-year-old girls. Along the outer perimeter of the blast, a number of others were knocked senseless by the strike, including the old one. A large chunk of tree had struck her, shattering one arm and breaking her leg in two places.

Then, as suddenly as it began, the hail stopped. The rain lessened, and the remainder of the group crept out from under the trees to gather at the site of the catastrophe. The alpha was dead, as were most of the other dominant females, and no one knew what to do. One female felt something touching her leg and looked down. It was the old one trying to say something. The woman knelt down as several of the others gathered around; maybe the old one could tell them what to do.

"The trees," the old one mumbled, "the trees, stay away from the trees." Now she remembered that time long ago when she was a child and the females had sought shelter from the hard white stones under a tree. Now she remembered. "The trees, stay away from the trees." And then she died.

Of the twenty-three females in the Lakeside clan, nineteen had taken part in the foraging expedition. The other four had remained in camp, with one in the last stages of pregnancy and the other just having given birth. In one stroke, the clan lost nine of its higher-ranking females—six as a result of the lightning strike, and three more would die later of wounds from the splintered wood. Ten of the children had also died in the blast, and four more would die of wounds.

The two disasters had left one of the largest clans in the area with only four males, fourteen females, twenty children, and six young males from the middle fire. It would be a long time before the Lakeside clan would regain the strength it held before the elephant and the hail. And things were about to get worse for the females.

Death was common in the clan, especially among the children. Barely half of them would reach reproductive age. But among adults, it was less frequent, and never in the band's memory had so many died at one time. The power structure further down on the dominance ladder in the clan was less clear and linear. Patterns were still being established, and the problem revolved around two sets of three females. The two trios had bonded early and well, working together, foraging, helping to look after one another, and sleeping together. Both trios consisted of teenagers. Each had a strong but inexperienced leader and two stalwart companions. They were evenly matched and in small ways tested their respected capabilities against each other on a regular basis. Under normal circumstances, one or the other trio would acquiesce or one of the two would lose a member, thus tripping the balance of power in the other trio's direction.

To further complicate matters, the remaining eight females—four pairs—split evenly between the two trios. The clan, therefore, had two almost evenly divided groups of females with seven in each group. Leadership would take time to sort out, but Lakeside did not have the time, and the Black Rock males would make the decision for them.

Each morning, the two trios would start out for a foraging site, and the remaining pairs would have to decide whom to follow. This day only one pair of females went with one trio while the other pair remained in camp, one of the two having just given birth. The five females made their way out on the veld in the direction of the Black Rock clan. This posed an element of danger for the group, but the site was also an especially good foraging area. The trio decided to risk the danger in an effort to obtain more food. The whole clan had been on short rations for some time now. The four males had secured some meat but not a lot, and all the clan's bellies were extremely flat.

At the foraging site, the females quickly built a fire and began to search for food. Bulbs, tubers, insects, small game, anything was taken to fill an empty stomach. Even as the sun climbed up and over the top, the females continued to forage diligently. Pickings were good. Food, while not abundant, was sufficient to take the edge off hunger and even provide a bit of fullness to growling bellies.

It was the group's inexperience coupled with their successful food hunt that did them in. Absorbed in exploiting the food supply, the females failed to see the Black Rock males sneaking up on them. When they did see the danger, it was too late.

The six males were among the group in a flash, using their clubs to kill both children and adults. In less than two minutes, it was all over. The five females lay dead, four children under the age of three were killed, all the male children who didn't immediately flee were killed, and two older female girls were killed. Of the little foraging band of five females and thirteen children, only three young girls were left—two five-year-olds and one four-year-old. The Black Rock males had not planned this as a raiding party for females but were not about to turn down the opportunity when it occurred.

The first thing the Black Rock males did was go through the meager possessions of the dead females. Lakeside had access to much better stone than did Black Rock, so digging stones were hefted and examined; several were kept. Spears, however, were another matter; none were anywhere near as good as Black Rock weapons, and so were left. There was nothing else.

As quickly as they had arrived, the men left the Lakeside territory, taking the three females with them. Soon the males were back in camp, and the three girls were regulated to the women's fire. Four, five-, and six-year-olds were preferred females when the clan raided other clans for women. Younger girls still needed some assistance from their mothers, older females tried to run away, and even the younger ones sometimes made an attempt. For that reason, the whole clan left the Black Rock kopje and went to a site on the far side of their territory. It would be almost a year before they would return to the great black boulder on a hill overlooking the veld.

As a postscript, one of the five-year-old captives—a clear-eyed, calm, and intelligent individual, wise beyond her years—rose over time from a spot at the bottom of the dominance structure in the Black Rock clan to the alpha female. Her wise and careful reign over the females resulted in Black Rock's growth to one of the largest clans in the area. She was truly the daughter of an alpha, and she would remember when the hard white rocks came out of the sky to stay away from the trees.

This is the way we lived for hundreds of thousands of years. Life was mean, painful, and a constant struggle. The idea that living was easy for our ancestors is a delusion. The concept of wandering through life—plucking a fruit here, picking a melon there, sipping from a limpid pool, sleeping in a cool leaf-filled bed—without a care or worry or concern is a fantasy. Our existence was short, hard, and brutal. The only person an individual could count on was a bond mate. And even then there were severe limits on capabilities, on ability, and on knowledge. Hailstones are a perfect example. Rare enough that the circumstances of such a storm took many generations to understand, and in the meantime, many died through lack of knowing.

Certain behavior patterns had survival value and continued for hundreds of thousands if not millions of years. The goal was to survive and reproduce. Behavior that served these goals tended to be retained, while behavior that did not contribute died along with the individual. Kindness, except among bond mates and between mother and dependent child, had little survival value. Altruistic predisposition, except among partners, was selected against. Intentional sharing within the clan was rare and outside the group, literally impossible.

Strangers were feared, killed, or fled from, depending on circumstances. Territory was defended because hunter-gatherers are very poor stewards of resources, and large areas were needed to provide enough food for each individual within the clan. A group could not allow another group to poach on its territory, taking food that the resident clan needed to survive.

And dominance was an integral part of our existence. Dominance provided order and continuity of the life our ancestors lived. Dominant individuals were more likely to survive; they had first dibs on food, resting, and sleeping places. They were dominant because they were stronger, and they were stronger because they were dominant.

Dominance was important and thus selected for, for many reasons, but the most important one was reproductive. Again we must go back to a basic tenet of reproductive behavior. All females need to reproduce because that provides the greatest number of offspring to the species. But only one male need contribute—the maximum male, the strongest, the smartest, the best male. This is the optimum reproductive formula.

As we have seen, optimum is not actual, but organisms that strive for optimum improve survival potential. We were animals. We are animals, and when it came to reproduction, we behaved accordingly—but not quite.

When a female of the clan became sexually receptive, she would solicit sex from all the adult males. The process was initiated by the female for a very good reason. Chimp females exhibit a swelling of the genital area when they are ready to mate. Male chimps are attracted to this visual display and mate with the females. In our case, the female genital area was hidden between the legs. In addition, our past habitation not only removed the visual aspects of sexual readiness but also the olfactory component.

What's a girl to do? Quite simply, when the female was sexually receptive, she came up to the male, tapped him on the shoulder, and said, "Let's do it." She knew; her body told her now is the time. And she knew what to do about it. She could be alluring, she could be coy, she could be brazen, and she could be wanton, but however she acted, she was ready to mate, and the males would oblige.

I have often wondered why it is that the human animal has turned sexual selection on its head. In all other social animals, it is the male that is attractive. It is the male lion that sports the majestic mane. It is the bull elk that carries the huge rack. It is the male peafowl that has the magnificent tail. And it is the male birds of paradise that are flamboyant.

Why is it then that the female human is the one that must be attractive, dolled up, pretty? Why is it that the female is put on display rather than the male? The answer is quite simple: over the eons, the female has lost her two most important signs of sexual readiness—sight and odor. The male would not and could not know when she was ready to mate, so she told him so.

Because her period of conceptual potential lasted only four or five days, the female would solicit the male when she came into heat. She would mate with all the males within that period, but at that point when she was most likely to conceive, she would often go off with the alpha or beta male and spend an afternoon or so in dalliance with one of these more dominant males. It was in her best interest and in the best interest of her offspring that it be sired by the most dominant male.

We see this predisposition today in female behavior. It is the female that dresses up, it is the female that paints her face, and it is the female that must have the right shape, the large breasts, the sexual face, and the come-hither look. And so females respond to deep and primordial urges. They also like guys with money.

It is necessary to be realistic about our ancestral sexual habits. Sex didn't happen very often. Let us go back to Black Rock clan of nine males and fourteen females. How often did a female come into heat? On average, four of the fourteen females would be likely to be available in any one year. This is based on a nine-month pregnancy and a period of lactation of approximately four years. This four-year period is reduced to an average of two and half years by the high infant mortality rate. On average, sexual activity took place four times a year—once every three months, maybe four or five days of high jinks, and excitement every ninety days. This doesn't sound like much, and it wasn't. Sex was important, necessary even, but not the big thing we would make of it.

Our ancestors had plenty of exciting things happen to them. Every day was a hazard, every trip out to take a crap an adventure, every hunt something to get the adrenaline flowing. We did not need sexual stimulation to give our days excitement or our nights bliss or our lives meaning. So in spite of the fact that some make much of our so-called continuous sexual availability, sex really comprised only a small part of our daily lives.

We like to think of ourselves as sexy animals, but we are not. We like to think of ourselves as monogamous, but we are not. We like to think of ourselves as kind, reasonable, and gentle. Many of us are, but some of us are not.

We do not like to think of ourselves as mean, nasty, and self-serving. We like to think of our species as something different, unique, not like other animals. But animals we were for hundreds of thousands, if not millions, of years. Just like in nature today, there was no justice. Just like other animals today, there was no kindness to the stranger and only limited amounts of it within the group. Just like other animals today, altruism did not exist. Is it any wonder then that some of us today act just like animals?

CHAPTER 9

This is a book of whys. The world is filled with pain and strife. We need to know why. Dictatorships were the only form of rule for thousands of years, and still today, many find themselves under the bootheel of some petty two-bit tin-pot dictator. And we need to know why. Like dictators, war has been with us always, wasting vast armies of young men along with tremendous amounts of treasure, and we need to know why. Capitalism works, and socialism does not work. We need to know why. Every culture that has ever been studied turns out to be a blueprint for status and dominance within that culture. We need to know why that is. On an individual level, homosexuality—a genetic predisposition—exists in the human animal, and it would be really, really nice to know why.

Sex is not as smooth as we would like it to be. We believe that males are males and do "male things." Females do only "female things." Anything else is considered aberrant and pathological. In spite of observation after observation of the contrary, many zoologists cannot or will not bring themselves to admit that homosexual behavior exists in most, if not all, animals. Their failure to confront reality is based on a delusion.

Homosexual behavior is common in the natural world. The human animal does not have a lock on same-sex sex. Let us begin with chimps, our closest primate cousins. Chimps regularly engage in sexual activity with others of the same sex. Bump rump is a form of simultaneous sexual stimulation common among female chimps. Two females—on all fours and facing in opposite directions—will rub their rumps together, stimulating their genitals.

Both male and female chimps engage in cunnilingus and fellatio. It is common, ordinary, and evidently pleasurable. It is also common that almost a third of all mounting activity among males is with other males. It's fun, it feels good, and chimps have none of the societal or cultural restraints that some human cultures do. Homosexual behavior is alive and well among our closest relatives.

But it is not just the chimps. Homosexual behavior runs rampant among the monkeys. Gibbons do it, and quite often, it is incestuous between father and son. Langurs do it—females acting as males with other females engage in the jumping display that precedes or follows heterosexual sexual behavior. Among rhesus monkeys, homosexual behavior can involve up to 90 percent of the females, although in some troops it is somewhat lower. Stumptailed, bonnet, and crab-eating macaque all engage in a host of homosexual behaviors. This includes males and females, young and old, and individuals from all positions within the dominance hierarchy. Homosexual behavior is common in primates.

Homosexual behavior is not limited to our wild cousins. Among marine mammals, dolphins—both the fresh and saltwater variety—engage in same-sex sexual activity. Orcas, fierce killers of the sea, like nothing better than a romp with others of the same sex, especially the males. Sperm and beluga whales have been observed engaging in homosexual behavior. Seals, sea lions, elephant seals, and walruses have all been studied, and all exhibit instances of homosexual behavior.

Such behavior has also been documented among hoofed animals. Deer, elk, caribou, moose, giraffe, antelope, gazelle, wild sheep, goat, and buffalo have all demonstrated instances of homosexual activity. Even the fat-tailed dunnart and the northern quoll do it.

Homosexual behavior is also common among birds. Konrad Lorenz studied greylag geese for years. Among his discoveries, Lorenz documented infant imprinting. Imprinting occurs in geese when newly hatched goslings imprint on the first large moving object they see. Normally this would be the female goose, but Lorenz got greylag geese goslings to imprint upon his lab assistants, thus permitting them to act as surrogate mothers in his studies.

One of the behavioral aspects of greylag geese that Lorenz observed was homosexual pair-bonds formed between male greylags. These pair-bonds exhibited all the behavior patterns of heterosexual pair-bonds and lasted just as long. In some cases, the male-male pair-bond endured for fifteen years and more. The only thing that the male pair-bonds didn't do is raise goslings.

Homosexuality is common among geese, swans, ducks, guillemots, albatross, grebes, and most aquatic birds. It also occurs in wading birds. Egrets, herons, and moorhens all show indications of homosexuality with both male and female homosexual pair-bonds.

As Konrad Lorenz's greylag geese demonstrate, sex is not the nice, clear-cut, simple genetic predisposition we would like to think it is. Male predisposition is primarily for females, but—and it's a big but—other males may also be on the list. So, too, with females; mostly they mate with males, but not always, and the species can survive with this type of behavior. Indeed, it seems that among the birds and the apes as well as most other vertebrates including reptiles, homosexuality is common. We should not be surprised that it occurs in our species.

It is most unfortunate that homosexuals are designated as being homosexual by the word *homosexual*. In the first place, the appellation refers to both males and females, so the prefix *homo* is incorrect. Secondly, the suffix *sexual* implies sex, and the origin of homosexuality in humans had nothing to do with sex whatsoever.

Let us go back to the question asked at the beginning of this book. Why is there a genetic predisposition toward homosexuality in a sizable minority of the world's population when, today, homosexuals have few, if any, children? If this present-day behavior were representative of our entire history, then homosexual predisposition would have long since disappeared. Yet homosexuality exists and has existed as far back into history as we would care to go. It occurs across as many cultural lines as we wish to study. And the question must be why.

Homosexuality is treated differently in different cultures. The Greeks accepted and countenanced homosexual behavior, affording it standing and respectability. The West, today, for the most part, grudgingly accepts the condition. Although it was only forty or fifty years ago that homosexuals were culturally, as well as legally, discriminated against by most of the Western world. But the real question is why it exists at all, and for the answer, we must go back to our prehistoric past, to our hunter-gatherer life on the plains of Africa.

In the beginning, homosexual predisposition was not the exclusive providence of the men, and it had nothing to do with sex. As we have seen, our place in nature's way of developing species was precarious. We had little in the way of weapons, no speed to speak of, and we were actually quite ugly. From an animal standpoint, being bare-assed naked, we stood out like a neon light—"Here we are, come and eat us." The future was not promising.

We did have one advantage, however—it was our evolving brain, our smarts, our intelligence. And because that is all we had, that is what we used. We made weapons using intelligence, we conquered fire using intelligence, we developed hunting strategies using intelligence, and we developed relationships using intelligence.

There is an old Indian adage that says two Indians together under two blankets will be warmer than two Indians each under their own blanket. The saying has an element of truth to it. Two Indians can share the warmth of each other, and two blankets provide more heat retention than one. It is, in the loosest sense, called cooperation. In our hominid past, it was called commitment.

We formed strong, lasting, and emotional bonds with our peers. We spent all our time with one or two others of the same sex. We slept together (in the real sense of sleep), we hunted together, we ate together, and we looked out for each other. We did so for the quite simple reason that individuals who engaged in such activity substantially improved their survival potential. We watched each other's backs, we looked after each other, we got to a point where we knew each other intimately but not "intimately." We were altruistic, nice, kind, and considerate, but only to our bond mates. We cooperated within the larger group as hunters and as gatherers, but our prime obligation was to our companion or companions. Alone on the plains of Africa, we were dead meat. Together and in small groups, we conquered the world.

The genetic predisposition toward forming strong emotional bonds with others of the same sex applied to both sexes. It was probably stronger in females than in males, although both lived hazardous, dangerous, and violent lives. It was stronger in females because they bore, by far, the most perilous part of procreation—the birth of a child. The male part of sex was easy. All he had to do was provide a tiny dab of sperm. The female had to push out a large helpless miniature replica of one or the other. The sex act was easy, a pleasure even, but birth was hard, exhausting, and painful. It is possible that several females would assist in delivery, but almost certainly one or two very close females with strong emotional attachments to the expectant mother provided the greatest assistance during birthing.

And so selection was for those individuals who could form strong emotional bonds with others of the same sex. To be able to attract and to reciprocate such strong emotional bonds improved survival potential. But the bonds were not sexual in nature. Today, for most homosexuals, they are sexual, but this is behavior, not predisposition.

Homosexual predisposition is a genetic behavioral propensity toward forming strong emotional bonds with individuals of the same sex. When the counting is finally done, I think we will find fifteen to twenty genes impacting this predisposition. This genetic makeup can cause behavioral patterns ranging from individuals who have strong bonds of friendship with others of the same sex all the way to committed same-sex pair-bonds.

It is my contention that what we refer to as homosexual predisposition is natural and normal behavior for our species. When all the phobias and biases are laid to rest, we will come to find out that to form strong emotional bonds to one or two individuals of the same gender is nothing more than a very successful ancestral survival mechanism.

It is also almost a certainty that our so-called male-female pair-bond is simply a misfiring of this genetic predisposition. We will deal with why this male-female pseudo–pair-bond exists a bit later. Suffice to say, the male-female pair-bond is a delusion.

We see acts of kindness, altruistic behavior, and sacrifice in the human animal, and we marvel that such a creature as ourselves could ever have been created. We see ourselves as the culmination of evolution, a fitting climax to natural selection, but we are not. Our caring comes not from some great plan but rather from simple survival. This behavior was originally reserved for our partner, but it, like pair-bonding under the dilution of surplus population, has also misfired.

But we are not all homosexuals, although the term can be used to designate a range of behaviors. Genetic predisposition influences relationships all the way from strong, lasting friendships to true same-sex pair-bonding. Using such a criteria, many more relationships can be placed under this genetic umbrella. This is not to belittle or impugn friendships, but rather to show that such relationships can have genetic roots. Still, we are not all the same, and the reason is surplus population.

Some individuals are as meek as toast, some as hard as nails. Some individuals have, what is judged to be in our culture, high moral values. These people are honest, fair, and forgiving. They are like most of us would like all of us to be.

But not all of us are honest, fair, or forgiving. We all know mean, nasty people who think only of themselves. Individuals who put their own desire and needs first with no consideration of others. These individuals make up a small but sizable proportion of our population.

Good people, nurturists say, are easy to explain. They are humane because our evolution has been the evolution of a humane animal. We are, according to the nurturists, nice. And yet it is obvious that we are not all nice. Some of us are nasty, and that needs to be explained. According to the nurturists, the explanation lies in upbringing. The problem is that these individuals were not treated right during their formative years. Had they been treated nicely, they would have been nice. But would they?

For many thousands of years, our ancestors lived one kind of life. We were scavengers, gatherers, and, to a certain extent, hunters. Life was hard, short, and very ugly. At first, we were inefficient predators, we were slow, we were puny, and we were marginal. We had only two advantages. We had fire, and every other animal on the plains of Africa feared fire. We also had the first glimmerings of intelligence. This made it possible for us to use fire both as an offensive and defensive weapon. We survived but not in very great numbers and probably not as a major contributor to predatory life on the African veld.

There is little evidence that our remote ancestors hunted, but some that indicate we scavenged. But even scavenging can be dangerous, as carnivores do not like to see others taking food they risk life and limb to kill. We probably did it with fire and spears. Lions fear fire, and a lion with a full belly, faced with burning branches and a few pointed sticks, may just opt for retirement. No matter, we survived and multiplied and slowly crept out of Africa and into the Middle East. We continued to survive and procreate, and in a hundred thousand years or so, we pretty much occupied most of the suitable lands of Europe, Africa, and Asia.

This was our first exodus out of the land of our evolution. It occurred roughly two million years ago. We were not much more than bipedal apes with slightly larger brains. For over a million and a half years, *Homo erectus* survived and scavenged and hunted the known world. But his reign was about to end.

Approximately two hundred thousand years ago, a new *Homo* left Africa. Essentially modern man in both physical size and brain capacity, *Homo sapiens* took the world away from *Homo erectus*. *Homo sapiens*, too, begat and survived and proliferated until the known world was occupied.

In the process, we learned. By 160,000 years ago, we had gone from scavenger-gatherer-hunters to hunter-gatherers with emphasis on hunting or gathering, depending on circumstances.

In Europe, much of the continent lay under vast sheets of ice. Much of what was left was tundra, not much plant food that we could use, but a habitat suited to grazing animals. Grazing animals are packed with protein and imminently usable by bipedal predators. And so we learned and survived and multiplied until we ran out of room. Essentially modern humans occupied every suitable habitat on Earth, save North and South America mainly on account of a Cro-Magnon Columbus had not yet shown up.

Most animals, both prey and predator, come into balance with their habitat. We do not see a sea of zebra on the African plain, nor gnu nor elephant nor even lion nor leopard. Much of predator predation comes from other predators both within and without the species. Lions will kill hyenas when they can, and any predator will kill the young of another predator if it finds them.

Prey animals, too, have *built-in* population controls. Rabbits go through cycles, and populations build year after year until the ecosystem can no longer sustain the population. All rabbits then become less healthy, and a disease that could be thrown off by the bulk of the animals under normal conditions wreaks havoc on stressed-out rabbits, and they die in their thousands. Lemmings too march to the sea when population levels reach critical mass. But it isn't really the sea they are marching to; the sea just happens to be in the way of where they want to go.

Nature finds balances for animals that use speed and claws and fangs and camouflage. But a glimmering of intelligence, a bit of fire, and a pointy stick can throw the scales out of whack. This is exactly what happened.

It happened in three places, and it happened more or less simultaneously in the Middle East, in Central Asia, and in Central America. This was the first signs of civilization and surplus population.

Why would we give up the freewheeling, exciting, hunting life for one of farming drudgery? We really had no choice. As we got better at hunting, we took a greater and greater toll of the native animal population. Game became harder and harder to find. This, too, enhanced our mental development as better and better ways of hunting evolved. We invented the atlatl or spear thrower and the bow and arrow. In the end, all this did was further reduce the amount of game available, making hunting less and less viable. Something had to be done.

By this time, we were more or less rational in the sense that intellect meant more than instinct. We already gathered, and one of our gatherings was the wild grasses, which were prolific enough that they could be profitably harvested in spite of their small size. We also had wild goats taken as babies in the hunt and raised by the very first hunter-gatherers turned farmers-herdsmen. It wasn't that we wanted to be farmers, and it is probable that many social groups, when faced with a choice, opted for pulling up stakes and trying to find better hunting grounds. The ones that couldn't do this were stuck—farm or die, herd or starve. And so they did.

The single most fundamental fact of any civilization is that the masses eat grasses. Hunter-gatherer groups cannot provide enough food to supply anyone with sustenance except themselves. Oh, there was the occasional shaman or medicine man, one or two individuals too old to hunt or gather but valued for their wisdom or knowledge, but hunting-gathering is very inefficient and prone to extreme fluctuation. A good year and game is plentiful, the rains come on time and in quantity, the land cooperates, and the green grass grows. The prey animals eat well, calve, and raise lots of young. The predator eats well, and under these circumstances, so did our ancestors. But only one bad year or two and predatory animals would feel the pinch. So would we, and we starved in substantial numbers many times.

But farming-herding was different. If the rains failed, crops could be irrigated. If much of the wild game succumbed to the drought, domesticated animals could be taken from one area to another in order to forage. Yes, it was hard work—from dawn to dusk—and drudgery every day. No excitement of the hunt, no gorging on liver and brain, rump and shoulder, kidney and tenderloin—that delicious sensation of gluttony offered by a full belly. Just daily gruel, not necessarily good or even good for you, but there day after day. This was our first security blanket.

And it all depended on grasses—wheat in the Middle East, rice in Asia, and corn in Mesoamerica. Grass is the key. Large numbers of people cannot exist under hunter-gatherer systems. There is just not enough food. But with farming, we took the first steps toward civilization and surplus population. It wasn't easy; farming requires a whole new mindset. It also requires a whole raft of new skills and social constructs. First, the mindset. Hunters hunted when hungry. Farmers farmed all the time. Hunters feasted and went hungry. Farmers could generally find something to eat at almost any time. Hunters roamed their territory, going from place to place as animals and other considerations dictated. Farmers stayed in one spot. Hunters made temporary shelter as conditions demanded. Farmers formed settlements, building wattle-and-daub huts among their crops and along ridges above cultivated fields.

All this required a different way of thinking, of doing, of planning. We became farmers because there was nothing else. We didn't want to; hunting was fun, exciting, and provided an adrenaline rush that was almost narcotic in effect. And this is why it took us tens of thousands of years to go from a hunter-gatherer lifestyle to a farmer-herder one.

And it was a tremendous change. The number one rule of farmers is that you don't eat the seed stock. At first, the gatherers just harvested the wild grasses and in the process dislodged enough seed to provide for next year's crop. Then gathering became more efficient, and it became necessary to differentiate between food and seed stocks. This must have presented our ancestors with many a dilemma. It's the middle of winter; grain stocks are running low. There is the seed stock, but if it is used, then there is nothing to plant; we will all starve, but later. But if we don't eat the seed stock now, we will starve now. And that's a dilemma. Hard decisions, hard times, but we made it through.

One of the first things farming brought into being was the widespread use of pottery. It is possible that our hunter-gatherer ancestors made pottery, but this is doubtful. Pottery is heavy, and most nomadic societies must carefully ration all portables because they had no pack animals.

In all probability, the storage of grain in pottery started with unfired clay jugs. Skins were tried, but moisture got through and molded the grain. Straw and grass baskets were tried, but rodents got into them. Stone vessels worked, but making them was hard and, as a result, they were not very big. Unfired clay jugs would work. Thoroughly dried and stored under cover, they kept vermin out and grain in. But they were delicate, easily cracked, and if left in the damp, they absorbed moisture, molding the grain.

Then something happened; perhaps a storage shed caught fire. Perhaps broken pottery was used to make a hearth or was burned in a fire. Anyway, someone, a genius perhaps, noticed that fired clay was much harder and stronger than unfired clay. And pottery was born.

But making pottery is a whole different concept from farming-herding. Farming is hard work but fairly routine for our ancestors. Herding is easy if dull, but pottery is neither simple nor easy. First of all, only clay will do, and then only certain kinds of clay. It has to be gathered, conditioned, made into pottery, and then fired. A new kind of thought process, a new behavior pattern was needed. At first, each farmer made his or her own pottery. But a good farmer was not necessarily a good potter. Before pottery, both males and females could do just about all the tasks required to survive. Some may have been marginally better at farming or basket making or caring for livestock, but all had to be pretty good at just about all the necessary tasks; survival depended on it.

Now one could be a not-so-good farmer and still survive, provided one could make pots. Selection was not always for strong, hard-bodied individuals with lots of stamina, but also for individuals who could judge the quality of the clay, recognize the right amount of water required to temper the clay and the right temperature of the fire that baked the pots. So we have the first rudimentary steps on the road to division of labor and civilization.

But how is a potter to get along? He needs to eat, yet he doesn't farm. He needs skins to clothe himself, yet he doesn't herd. And so barter rears its economic head. The potter agrees to make pots for the farmer in return for grain. The tanner agrees to tan skins in return for grain. This is how it started—with pots and hides and grain.

This is how civilization was born, and with civilization came surplus population. Surplus population can be defined as that number of individuals a given area can support over and above what that same area can support as hunter-gatherers.

For over 160,000 years, our ancestors were selected for certain basic characteristics. They had to have good eyesight, good hearing, good toolmaking ability, be able to plan hunts with others, make fire, and, most importantly, defend the territory. With civilization, this changed. We no longer had to be as strong. True, some of us had to be, but less robust individuals could survive and reproduce. Individuals whose eyes were not quite as good could still make a contribution. We did not all have to have the traits of the typical hunter to survive.

The selection process was no longer the same, and just as it was for physical traits, so, too, the selection for behavioral traits changed. As we have seen, hunting-gathering is different from farming as farming is different from pottery making.

At one time, we were animals. For millions of years, we behaved just like animals. We were ruthless. The strong took from the less strong without thought. It just was done. Death was common both inside and outside the clan. Fights occurred; clubs and spears and teardrop-shaped stones were used both for offense and defense. Men and women died as a result. But there was no justice; there was no retribution, no stigma attached to the killer. It was just something that happened.

Among the !Kung San of the Kalahari Desert in Africa, murder is rare. Indeed, it is so rare that the !Kung San were initially passed off as peace loving, passive, and nonviolent. One anthropologist, who should have known better, claimed that the !Kung San was the original affluent society. Plenty of food, he said. Little work required to get by and very peaceable.

Further study demolished this claim. The murder rate among the !Kung San is as high as or higher than the murder rate in New York City. The !Kung San kill each other on a regular basis, but because population densities among the !Kung San are so low, it took years of observation to uncover the facts. !Kung San's murder rate rivals New York's as a percentage of their population and violence among these so-called peaceful hunters is much higher than in any civilized society.

But murder among the !Kung San did not result in justice or punishment or retribution. In later times, the offender was required to make some contribution to the relatives of the victim, but for thousands of years, dead was dead, and the living must go on living.

So, too, was life among our ancestors. Anger led to violence, violence sometimes led to death, and the killer always lived longer than the victim. Genes for violence persevered. Genes for taking were retained, and a predisposition toward self-satisfaction prevailed. We were animals for millions of years—selected for, maintained, and passed on because such selection worked.

Then things changed, and surplus population came into being. Now less violent individuals could survive and reproduce. Less assertive individuals could contribute to the gene pool. This diluted the very narrowly focused hunter-gatherer behavior pattern. Soon, less dominant men—and women—made up a majority of the population. Soon, people began to think that less assertive was normal and that the takers were atypical. A whole new concept of human behavior came into being. We were nice, we were kind, we were altruistic, and most of us were.

There is yet another aspect of our behavior that contributed to a radical change in human behavior, and that was war. Not war itself, which will be dealt with later, but rather the effect of war on surplus population. War is not good, but it does have one beneficial effect on populations. It was the more aggressive individuals who went to war. It was, on average, the more aggressive individuals who were killed. This had the effect of lowering the overall genetic predisposition toward aggression among populations. Down through the years, it has been the selecting out of more aggressive individuals on a regular basis that has gone a long way toward our present passive predisposition makeup.

But there is still a genetic minority that reflects our hunter-gatherer past. The taking, the violence, the killing—all of it is there. And while surplus population could dilute the gene pool, it could not expunge the violence or killing.

Surplus population is not taken into consideration in any nature/nurture studies. All individuals are assumed to be the same, to have the same genetic makeup, to behave the same way under similar conditions. And this just isn't true.

Any study of human nature must be based on two fundamental parameters. First of all, we are evolved animals. We come with all the genetic baggage of our evolutionary past. It is not good, it is not pretty, but then neither is nature. Secondly, we are not all the same. We do not all have the same genetic package. We differ, and sometimes that difference is naturally more aligned with the habits and behaviors of an animal. We are animals. And it should come as no surprise that some of us act as if we are animals.

The theory of dominance and delusion is not just some more or less arcane concept among many such obscure ideas. Dominance and delusion play an important part in the everyday lives of individuals, groups, and nations. The theory can be used as a blueprint, as we shall see, to improve, through understanding, the lives of everyone on Earth. But first, a disclaimer:

Some individuals and groups will not fare well in this analysis. Statements will be made that are not politically correct. Cultures will be examined and compared. Many will be found wanting. Races will be looked at, not always in a benign manner, and judgments made that are not always beneficial. This is, however, not a racist book. With a very few exceptions, I do not like anybody very much, and I do not discriminate in my dislike. We are human and, when viewed dispassionately, not a very nice species. It is true, as I have said, that only humans can and do commit inhuman acts. We have, down through the years, committed some truly inhuman acts, and we must understand why we act this way if we are to truly comprehend human nature.

The problem lies in the fact that we view these acts as inhumane when, as has been shown, they are very human, natural, logical, and even ordinary. And this is the rub: we want to think of ourselves as nice, kind, gentle, altruistic, and well mannered. Unfortunately, the people who count—the important ones, the alphas—are not, and this must be explored and explained. The answers are not nice, and nice people (there are a great many of them) will not like the conclusions drawn. Dominance and delusion characterize the human animal. The world today is defined by these two fundamental attributes, and the reasons why lie buried in our evolutionary past.

Why do people differ? Why are some nice and some not so nice? Why are some domineering and authoritative while others are shy and retiring? It is nurture, we are told, how the child is brought up. There is some truth to this concept, but much of what we do and say and how we behave is influenced by genetic predisposition. If the genetic predisposition is strong enough, no amount of "good" upbringing is going to change the individual. The question is why is there so much variation in the human animal? To find out, we must first go to the dogs.

As we have seen, breeding has managed to pull out of the basic wolf stock a literally amazing amount of variation. Big dogs, small dogs, dogs with all kinds of coat colors and hairstyles, dogs that herd, dogs that pack, dogs that pull, dogs that guide, and dogs that lap—all from the basic wolf stock. But our breeding of dogs from wolves has also produced animals with a wide range of temperament. Wolves are vicious, mean, nasty animals, and they have to be to survive. And yet we have direct from wolf ancestry cocker spaniels—brown-eyed, silky furred, and nice. There are setters of every stripe and kind, elegant, devoted, and a hunter's delight. Black Labs are probably one of the gentlest dogs around. On the other hand, we have German shepherds, good police dogs, enough said, or mastiffs, two hundred pounds of "don't touch me because I bite." All these behavior patterns and more. We have shy dogs, extrovert dogs, dignified dogs, and even classic clown dogs. All from a basic wolf stock. So many genes, so little expression in the wild.

Nature calls for certain basic characteristics and ruthlessly selects out any deviation from the standard plan. As a result, foxes and jackals, and coyotes and wolves all share the same uniform body plan and basic colors. The variation is only in size, and that variation is dependent upon habitat and resources. Natural selection picks and chooses. Wolves with the short legs of a cocker spaniel would not be able to keep up and would be selected out. A wolf with little or no dominance predisposition will not mate or have offspring, as only the dominant animals will breed. The list goes on and on. Only certain specific and very limited attributes are chosen for because these characteristics have survival potential. But the genetic potential for a great deal of variation still remains in the basic wolf stock. So, too, it is with the human animal.

For hundreds of thousands, if not millions, of years, our species, too, was chosen for certain very specific physical and behavioral characteristics. We were as mean, as nasty, as belligerent, and as combative as any other predator on the plains of Africa. It was a strategy that paid off, and we became the only large carnivore species to populate the entire world. Our ancestors were not nice—tea and crumpets not on the menu—and altruism was definitely not in our behavioral repertoire. There was one exception. One, or perhaps two, individuals toward which another individual might show compassion. So all is not lost, just badly scrambled, and again, the reason why is surplus population.

As we have seen, surplus population occurs when a group of individuals figure out a way whereby the territory is made to support more individuals than can subsist on that same territory by simply hunting and gathering. Hunting and gathering are very inefficient survival mechanisms. When fruit is ripe, there is much more available than the group can eat. When meat is slaughtered, it is immediately consumed because there is no way to preserve it. A clan, in order to survive, must have a fairly large territory but cannot harvest all the assets that are within that territory.

Territorial potential and usage is critical to our understanding of how our ancestors evolved from hunter-gatherers to farmers-herders. The territory only has a limited potential with respect to population densities, and our ancestors were quick to reach this potential. In places, this was the limit of human potential given the capabilities of the people and the type of habitat. It is only where circumstances and potential both occur that we were able to cross the divide from hunter-gatherers to farmer-herders. We need look no further than the first settling of North and South America to see how the process works.

Fourteen to sixteen thousand years ago, a group of individuals crossed the land bridge between Asia and North America. Some say they came by land, some say by sea; some say it was much earlier than that. Some say they came from the east, but this is not really important. Within, at most, two or three thousand years, these immigrants had populated the entire North and South America landmass. From no more than a few thousand, these people blossomed into a population numbering five to ten million—a truly prodigious feat.

They had help. As the new predator on the block, prey animals did not know and fear them. As a result, it was fairly easy for these hunter-gatherers to approach prey. There was even an incentive to keep moving, exploring, and seeking new hunting grounds. New places meant new prey animals—animals that had no previous contact with humans and did not know how dangerous they were.

When every nook and cranny of the Americas was occupied, every territory settled, the AmerInd's population densities stabilized and, with three exceptions, remained relatively constant for the next ten thousand years.

How can this be? The potential to reproduce was still there. Certainly females bore as many children as before. Well, maybe not quite. If we, at some point, reached equilibrium with the food supply, then population levels would stabilize. But this is not the case. Huge herds of buffalo roamed the plains of North America. The eastern woodlands were full of deer, elk, bear, and even an eastern woodland buffalo. Like most hunter-gatherers, we did not eat ourselves out of house and home.

Tribes staked out claims to territory and defended them. They claimed large territories because with such extensive holdings, game was easier to obtain; fruit was more likely to be available; and roots, shoots, and tubers were easier to find. And we protected our resources. Warriors raided neighboring tribes, killing or enslaving only to be raided, killed, or enslaved in turn. The history of the people of North and South America prior to 1492 is a history of war, killing, and pillage. This is what kept populations stable in the ten thousand years before the coming of the Europeans. This is how we lived. This is how we survived. The strong and the able just barely kept population densities constant. Less able and weaker social units were selected out on a regular basis. Nice did not survive; nasty did.

There were three exceptions, but they, too, demonstrate dominance and delusion as well as surplus population. The Inca in western South America, the Aztec in Central America, and the Mound Builders in the Mississippi River valleys managed in the very last of that ten-thousand-year period to produce surplus population. These surplus populations were built upon corn.

Corn is a naturally occurring grass growing in the highlands of Central America. Today, one can still find these primitive grasses growing in remote spots of Central America. Corn, unlike wheat, was initially not a very good grass for building surplus populations. The cobs were small; the kernels were tiny and difficult to harvest. The grain was seasonally gathered to supplement other foodstuffs including squash, beans, and peppers. As long as wild corn was the only corn available, population densities remained low. But when special attention came to be applied to the grain, when corn was selected for and cared for, it became a truly life-sustaining crop. The surplus population boomlet of North and South America was founded on corn. It occurred because the masses must eat grasses, and dominance, already in place, bloomed in its wake.

Looking at the history of the Inca and the Aztec, we see dominance writ large. Emperors and kings were in charge, a dominance hierarchy, the vast bulk of the population under the thumb of the rich and powerful. We also see the ritual murders of thousands, if not millions, of individuals in an attempt to satisfy some longing, some demand of society. Blood by the bucket and skulls by the score to gratify a myth or to delight a deity. These people went from simple hunter-gatherers to a fairly sophisticated civilization, but dominance and delusion persisted—a common characteristic of our species.

But the AmerInd experience in the western hemisphere does tell us something of importance. We, as a species, are capable of producing large numbers of offspring and do so under favorable circumstances. We can and do multiply like rabbits when the conditions permit. We produce offspring that might not otherwise survive, but do because there is unused territory into which less capable individuals can go. Once all the territory is occupied, we are forced to fight each other for a piece of the pie. Populations stabilize, and only the most ruthless are selected for.

But surplus population is the key, and this is what needs to be explored. Like the dogs we breed, when selective pressure is relaxed or even eliminated, all kinds of genetic predispositions appear. At one time, our species was selected for very specific responses. We needed to be brutal, mean, self-centered, and selfish. These predispositions worked. They made existence and survival possible so they were rigorously selected for.

But with surplus populations, things changed. Now almost everybody had a shot at survival. Maybe an individual was not so good a hunter, but he or she could do something else important. This allowed that individual to survive and, even more importantly, play a part in the procreative process. What is important here is that we were no longer selected for by certain very limited and very specific behavior patterns. People who were less able to hunt, less able to survive in small hunter-gatherer bands were now surviving.

Early evolutionists using Darwinian-inspired concepts decided that nature was all. We were evolved animals, responding like all animals to our genetic predispositions. Then nurturists like Margaret Mead came along and said, "No, it was not nature; it was nurture." How we were treated when we were young was the key. If we were treated well, we grew up to be productive, law-abiding citizens. If, on the other hand, we led a deprived childhood, we led a depraved adulthood. The key was to ensure that each child led a nice, kind, altruistic childhood. Unfortunately, both these theories are right, and both are wrong. The human animal cannot be compartmentalized. We are each different, each with an individual genetic load, each predisposed to behave in certain ways regardless of our upbringing. Yes, upbringing can influence behavior, but genetics are just as important in determining our behavior. Certain people will be predisposed to behave in certain ways regardless of upbringing. So both nature and nurture play a part in our behavior, and different combinations of both occur in different people. This is the effect of surplus population.

Cultures must also be considered. Cultures can and do impact how we behave. These factors influence each of us in different ways. How we are raised will have an effect on how we behave. Our individual personalities will impact our behavior. And genetic predisposition will have its own effect. All three define our behavior, and because personalities differ and cultures differ and genetic predispositions differ, we are, each of us, going to respond individually to circumstances. We cannot be treated as a whole but must be considered as individuals.

We have some individuals in our societies who are celibate. Now this is a real behavioral anomaly. Like many species, our goal for hundreds of thousands of years has been to survive and to reproduce. This is a very basic, very fundamental drive, which is deeply ingrained in human nature. And still some can resist this genetic inclination to the point that they do not reproduce. It is a behavior pattern that is culturally inspired, and it is unlikely to be genetically passed on.

These people have very strong control over their genetic predispositions. Most of us cannot or will not abstain from a reproductive genetic predisposition, and we form the vast bulk of humankind. There is a third category that we must consider, for much of both the good and the evil in man comes from this group.

While surplus population can and does produce nice, kind, altruistic individuals (lots of them), it also produces a small minority of individuals that carry the genetic predisposition of our hunter-gatherer ancestors—individuals that will do whatever they have to, whatever the culture will allow them to in order to achieve their dominance goals. Cultures can influence and direct these individuals, but first the culture must recognize and understand what makes these individuals act the way they do.

CHAPTER 10

Something happened forty thousand years ago. Something, someone, somehow changed, and the evolution of the human species dramatically altered direction. It has been described as a sociocultural big bang, a human revolution, and a creative explosion. What happened forty thousand years ago could also be described as the advent of delusion.

In chapter 2, I explored the idea that belief in a supernatural being might have been just enough of an advantage to a contending caveman to cause such a belief to be selected for. I shall argue that forty thousand years ago, we began to believe, and culture, religion, and delusion were born.

The dates fit pretty well. Forty thousand years ago, we were finally settling Australia. This is the last large land area close to the Afro, Euro, Asian landmass, and we now occupied all of it. The land from horizon to horizon was inhabited. We could no longer ease our population problems by expanding our range. Every continent was inhabited except North and South America. They were off-limits because of the glaciers in Alaska. So we were placed in direct competition with each other, and something had to give.

Prior to forty thousand years ago, our ancestors had maintained a very conservative lifestyle. Our species is identified by tools. Most were of wood or horn or antler and have long since disappeared. But some of what we worked with was stone, and stone endures. Our earlier stone tools were simple and not the easiest to identify, being simply cobbles with a chip or two knocked off. Still this tool industry remained our principal lithic contribution for half a million years. Then some genius developed a different way of "making" stone tools.

He or she wasn't too much of a genius though. The new handy-dandy toolmaking art consisted of preparing a rough core prior to striking a blow that would produce a bifacial tool. These were superior to cobbles, and as a result, they were the mark of man for over a million years. We remained a very conservative species, and so very little changed in the toolmaking province for a very long time.

About 140,000 years ago, we again benefited from a very smart individual reacting to environmental demands, and a new suite of tools came into existence. These tools were made with a degree of skill heretofore unseen in human history. Blades were made by the removal of very fine chips, leaving a tool that was not only superior to previous tools but was also beautiful to behold. This type of toolmaking lasted 100,000 years and was the precursor to everything that follows.

Improvements in tool technology that had served us well for two million years gave way to an explosion of innovation. Tool kits changed five times in the next thirty thousand years. Bows and arrows came on the scene. Throwing spears appeared along with atlatls. Cave painting developed with its artistic red, yellow, and black textures. The first ostrich eggshell beads were made. This is also the time that the enigmatic Venus figurines came into existence. Our first representations of ourselves, these figurines can best be described as female and fat. At a time when sustenance was neither sure nor adequate, a fat female was a sign of plenty.

But what was the change? What happened forty thousand years ago? This is only speculation and in all probability unprovable, but I think that at some time before the forty-thousand-year-old explosion, someone was born who was slightly different. Or maybe not so slightly.

We made and used tools, we made and used fire, and we populated the known world. We were by any measurement a successful species. I think that we were also conscious in the sense that we could visualize the future, grasp the concept of seasons, and make decisions based on input more than instinct. We could think and plan and execute with an awareness of what we were doing. But we had not the potential of modern humans because we lacked one genetically programmed ability. We had yet to be gifted with a creative faculty. We did not have the ability to imagine.

The ability to imagine predisposed itself in the human animal probably fifty thousand years ago or so. By forty thousand years ago, it had become prevalent enough in the fossil record that we begin to see the results.

There are a number of ways it could have come into being. Perhaps we always had the mental potential but lacked the synapses to access that potential. Perhaps it was a brain change, the genetic manipulation of gray matter that gave us the ability to imagine. Whatever it was, however it happened, it made us, us. One person over forty thousand years ago (perhaps there were more) changed the direction of human evolution and gave us the ability to imagine.

It probably started with animals. We were, as I have said, not very impressive predators. All large predators, and even some of the smaller ones, were considerably more admirable as hunters. Lions and bears and wolves were all much better equipped to kill than man. Perhaps it started by someone thinking, "What would a wolf do?" Or perhaps it was a comparison of strength. One individual of substantial size might compare himself to a bear. Or a particularly fast hunter might identify with the cheetah. The thinking became being, and so the clan of the lion or the clan of the wolf was born. In the process of identifying with a specific animal, we at first venerated the physical attributes of the predator and, as a direct extension, the spiritual protection of the animal as our clan totem.

It began with knowledge and delusion. The knowledge that somehow in some way, we had the support of the supernatural. Somehow in some way, if we believed strongly enough, we could prevail and pass on our predisposition to believe. Then it became more than just believing. We, as all of us came to believe, lost our basic believing advantage. And so we had to do more than just believe. Rituals were developed; certain objects obtained importance all out of proportion to their real worth to serve our myths. Behavior had to be just so, cave paintings necessary to build confidence. Dances and demonstrations were required not only to show the Holy One our devotion, but also to build up our own confidence in our ability to contest.

For many, to imagine is to believe. This makes sense from an evolutionary standpoint. At one time, to imagine that one was the recipient of supernatural benefits led directly to believing that one had the support of the supernatural. The slight advantage thus engendered was often enough, just enough, to permit a favorable outcome when contesting with others. Selection would be for those who were capable of equating imagining with believing.

Not only did we have to believe, but also that belief had to be maintained even if events did not always end favorably. A mechanism was selected, for that would cause otherwise rational, intelligent thoughtful individuals (and some not so) to hold concepts in their brain without the slightest rational basis. This was, perhaps, a one-way synapse or even a one-way neuron. Once the concept was in place, the brain stopped any connection with the concept. Outflow was allowed, and behavior predicated upon the beliefs was implemented. But because the genetically selected for belief preserver was placed off-limits to rational inquiries, we are able to believe the dumbest things without question.

And so religion was born, and culture was born, and delusion came into our lives. Categories of men (it's always the men) were born who would devote themselves to the care and maintenance of our religious beliefs. These individuals had to be maintained by the rest of the population. Thus was instituted a vicious feedback mechanism. The men who maintained the religious rituals were motivated to act in ways to improve their own position while at the same time claim that it was the religion that mandated their demands.

This is how religious authorities achieved the dominance they truly craved. It was all done in the name of God.

Once a program of religious practice is established, it is enforced through the development of cultural demands. Religious intrusion can infiltrate every element of the culture. How people act, what they say, even how they think are all bound up in religious mores. And so cultures are born.

Any group of individuals organized for a purpose can be defined as a culture. The defining component of all cultures is dominance. Sometimes this dominance is informal and casual, but real nonetheless. Sometimes it is rigid and authoritarian with extreme consequences for any lapse. Cultures have laws, rules, and regulations used by the individuals within the culture to govern and control. Cultures have rites and rituals, which, like the rules and regulations, delineate the limits of acceptable behavior. These rules and regulations also serve to mark the boundaries of the culture.

Cultures offer us a method of making sense and order out of what otherwise would be a chaotic mess. Each individual within the culture knows where he or she stands. They know, as well, what steps are needed to advance their relative position within the culture. They also know exactly what not to do within the culture. They know that certain acts or behaviors will cause them to lose status and perhaps be expelled or even killed. Cultures are necessary and important to man: they offer order, they blueprint behavior, and they present a road map to dominance within the social group.

Simple hunter-gatherer groups may have only one culture, or perhaps a tribal culture and a subculture for men or women or both. The group as a whole has a set of rules and regulations and rites that define that particular social unit. Then within the group, the men may have a whole subset of rules and regulations and rites that define the relative status of every male within the group. The females may also have a setup whereby their relative position within the group is defined. In other cultures, the female sometimes takes on the status of her mate. But it is always the same: believe this, do that, then that, then that. Do what is required well and be rewarded. Do it poorly or not at all and be expelled.

Cultures are also strong behavior modifiers. As we shall see, all cultures are dominance systems, and individuals within the culture will act in ways to improve their status even if it means forgoing reproductive potential.

During World War II, Japanese soldiers refused to surrender because of cultural prohibitions. As a result, many more were killed by the Allies than would have occurred had there not been a strong cultural abhorrence of surrender. These individuals gave up their lives because such actions were looked upon with approval by the Japanese culture.

Never mind that this action resulted in a situation where these same men would not be passing along their genes. They gladly gave up the two most important goals of all life—survive and multiply—in the interests of pleasing their cultural commandments. Cultural demands have a huge impact on human behavior.

Islamic suicide bombers also fall into this category. They lose the potential to reproduce, but they attain the status they crave within the cultural group. They are honored, they are praised, and they go to their deaths with the final thought of seventy-two virgins as their eternal reward.

Cultural demands often override genetic predispositions for self-preservation. At times, it becomes difficult to see which component—genes or culture—manifests itself in human behavior. We see individuals risking their lives to save others and claim an altruistic past when, in fact, we may be seeing culture. We observe sharing and claim sharing as a unique trait of the human animal. But many cultures do not share, especially with anyone outside of the culture. In today's world, nothing could be more important than to understand why people do the things they do. Nothing reflects this more than the way a culture views the taking of a life.

This varying value placed on life can be no better demonstrated than by two TV programs shown during the Afghan invasion. The first program was a *Law & Order* episode that dealt with the reaction of four cast members to the death by lethal injection of a convicted murderer. One of the cast, a recovered alcoholic, falls off the wagon. Another takes out her emotional trauma with a bout of extreme jogging. All are affected; all demonstrate a strong emotional response. Death isn't nice. Killing, even when approved by society, is not easy. In our culture, taking a life is emotionally traumatic.

At the same time, there was broadcast from Afghanistan a short newsreel showing the cruelty of the Taliban. The segment involved a female who, after enduring years of abuse by her husband, finally killed him. She, too, was sentenced to death. She was brought to the killing fields in the back of a pickup and made to kneel on the ground. The camera (hidden, of course) shows her with her hands tied behind her and her head bowed. To one side is a group of four men. All are armed with AK-47s, and they appear to be engaged in casual conversation. After a few moments, one of the men goes over to the woman, points his gun at her head, and pulls the trigger. The reality of the execution becomes clear when the viewer sees a spray of gravel kicked up by the bullet after it passes through the woman's head and hits the ground. The executioner turns back to his buddies, and they resume their conversation, totally oblivious to the dead woman.

In our culture, the sanctity of life is an important cultural value. In the view of the Taliban, life is cheap. In many Islamic cultures, life has little value as long as it is someone else's life. The cultural values of the Taliban and Islam, I fear, more closely represent the true view of our kind. Those cultures of kindness we hold so dear are delusional.

Some cultures have a strong sanctity of life component, and some do not. Here in the cultures of Western Europe and North America, we value life highly. A criminal convicted of murder and sentenced to death goes through fifteen to twenty years of appeals to ensure that the taking of a life is the correct course. We go to great efforts to "save" people on life support, and we have a very strong lobby that insists that "life" begins at conception. Taken altogether, our culture values life, and any attempt to deny someone a life is viewed with distaste. Individuals who appear to have little or no regard for life are not accorded status within our society. This is, I think, a good, healthy attitude, but it is not universal.

In most Islamic cultures, life is cheap. In most African cultures, life is cheap. In many Asian cultures, life is cheap. Until very recently, in most Central and South American countries, life was cheap. Our Western European and North American cultural value of life is a minority position. This is something we need to know and understand when we deal with other cultures. Sanctity of life is not a "natural" human predisposition but rather a cultural manifestation. To think otherwise is a delusion.

*****

In the United States, we have many cultures. We have two baseball cultures: one rigid and regulated, the other informal and relaxed. The sport of baseball, like all sports, is a sport of dominance. One team will be the final victor at the end of the season, having beaten the next best team in a seven-game series. Rules have been laid down, regulations promulgated, and boundaries set. Gloves must be just so. Balls must be of specific size and weight and make. And no corked bats, please. Plate size is defined, bases are described, and diamonds designated. The very minutiae of the game addressed, delineated, and determined. The sport is very rigid, very formal, and the team that follows the rules and wins is dominant over every other team within that particular baseball culture.

Then there is the more laid-back baseball culture. This is the culture of knowledge. The individual who can memorize the most statistics, recall the most players, and describe the most individual plays and games achieves status and accord within this less formal baseball culture.

All sports are dominance cultures. Baseball, football, basketball, tennis, horse racing, soccer, and on and on and on—all dominance systems. Win at all costs, but within bounds. Whatever it takes, whatever needs to be done, win, win, win—dominance writ large.

There are other cultures in the United States. Fraternal orders are cultures. The Elks, Moose, Rotary, Lions, Masons—all these and many more are all structured cultures whereby certain behavior is admired and rewarded and other behavior looked down upon and despised.

There is yet another subculture in the United States, which poses a singular problem for us. Juvenile gangs, especially male juvenile gangs, are dominance systems. They are hierarchies dedicated to establishing relative rank and position within a selected peer group. Such behavior is old beyond time and deeply ingrained in young men only recently introduced to civilization. This is one of the more difficult problems of today's world. Peer bonding among young males is normal, natural, and, at one time, had great survival value. We should not be surprised that it is such a strong element in some cultures today. To see why we have juvenile gangs today, we must look at juvenile life forty thousand years ago.

The springhare pauses, looks around, and sniffs the air. Something isn't right—but what? Crouched in the tall grass twenty yards away, a young predator watches the animal. He is well positioned, downwind, behind sparse but adequate cover, and, most importantly, next to the path the springhare should take. Both animals wait, one patiently, one with growing apprehension. Again the hare sniffs the air, pointing its nose this way and that, up and down—something just isn't right.

A few hesitant steps—half-hearted hops, really—and the hare stops again. The young predator's eyes glow with excitement and anticipation. His heart pounds loudly enough; he fears that his prey will hear. A few more steps and the springhare will be in range. The predator is young, but the prey is younger; a more experienced animal would trust its instincts and bound away immediately. But the springhare, in its immaturity, cannot make up its mind and takes a few more hesitant hops.

The predator stands up and takes one stride forward, at the same time casting a short wooden spear. The boy uses an atlatl, and he is good. Aiming at where he anticipates the animal will be, he launches his missile with deadly accuracy. The spear catches the springhare in midstride. It strikes just behind the rib cage, and the weapon goes right through the animal, a good foot of it sticking out of the other side. The springhare lets out a terrified bleat and collapses into the grass. The boy is on it almost immediately and, using a small club, kills the animal by smashing its skull.

For a few seconds, the young hunter stands over the hare, allowing the delightful thrill of the kill to wash over him. But this place is not safe, and many a hungry predator would take his prize and him, too, if he isn't careful. Looking back the way he came, he whistles twice—the sound the hyrax makes when danger threatens. Four heads pop up from behind a large boulder some fifty yards away. One quick glance around, and they run toward the whistler.

All five boys are young—the hunter, thirteen; two others, twelve and eleven, but closer than that being only about six months apart. The last two are both nine years old. All five are black, not brown, not chocolate, but a dark ebony black, so black that when they play in the water, they glisten. All five are thin, well muscled, and naked except for a hide belt around the waist and a small flap of skin in front. All carry three or four spears—thin rods of wood hardened and stiffened by careful application of heat, their tips drawn to a fine point by the removal of tiny curls of wood done with a sharp flint flake. Each also carries a small wooden club, fourteen to sixteen inches long and heavy for its size. It is made from a very dense wood, fashioned slowly one sliver at a time through countless hours in front of the evening fire. In a pouch hanging from the belt, each boy also carries a hand axe. It is made of flint, shaped like a teardrop, and big enough so that the big end fits comfortably into the palm of the hand. In addition, each has five or six stones—round, flat, and one and a half inches across. The little stones are the very first weapons the boys ever carried, and each is deadly accurate when throwing the tiny missiles.

The fifth boy, the youngest, has something else tied to his belt. It is a pair of clamshells—white, worn smooth from use, and tied together with bits of hide. The two halves fit tightly except in three or four places where the lip of one shell has been broken back, leaving a slight gap in the alignment. From these gaps, small wisps of smoke arise only to fade away on the breeze. The smell of wood smoke is constantly in the air.

The younger boys admire the hunter's kill, prodding it with their spears. The animal is limp, loose, like some half-liquid mass in a fur-covered sack. With a grunt, the hunter extracts his spear and checks his point for damage. He gives another grunt of satisfaction when he sees none and picks up the springhare. Slinging the carcass over his shoulder, he and the other four set off for Black Rock.

Five hundred yards from camp, the young hunters pause for a few minutes as they study the rock-covered hill. The boys have to be careful. The males, if they are at camp, might take the springhare. Taking is ordinary, if uncommon. Most of the group's possessions are only marginally worth taking, but it could happen. One day, one of the younger boys found a fine hand axe, no doubt dropped and lost by someone. It was made from a very close grain chocolate-colored stone that was actually semitransparent at some points along its margin. It also held a very fine edge, much sharper than the flint or quartz stones generally used for such purposes. He did not have it long.

It was quickly taken by the eldest boy at the middle fire. He just walked over to where the younger boy was using the tool to strip long, thin shavings from a new spear, and took it. For a minute, he admired the fine edge that someone's expert chipping had produced. Then with a grunt of satisfaction, he put the hand axe in his waist bag.

The eldest boy had the fine chocolate hand axe for all of a day before the dominant male commandeered it for himself. It would remain the property of the alpha until he lost it, until he died, or until he lost his dominant position. Taking was ordinary, expected, and sensible. The fine chocolate hand axe with its exquisite semi-transparent edge would be best used by an alpha.

The springhare faced no such danger, although a small antelope had been commandeered by the males only a few days earlier. The animal itself had been taken from a cheetah too winded after the chase to contest the taking. When the boys came back to the camp with the antelope, they had checked to see if the way was clear but had failed to see the dots on the horizon that heralded the arrival of the males from an unsuccessful hunt. The males, without a word upon reaching camp, took the prize.

This time, the boys scanned the veld carefully and, after assuring themselves that they were alone and would be for some time, hurried into camp. Face is important; it is good to be back at Black Rock, but the youngsters are loath to let anyone, especially the females, know. So five young boys, the eldest barely a teenager, casually saunter into the shelter. The springhare is placed aside as the boys turn to their fire. They drag up brush and wood; they stir the hot embers to get the flames to appear. The fire keeper gets little from the middle fire and does little to maintain it.

The boys gut and skin the animal and throw it on the flames. Soon, five hungry mouths make short work of the animal, consuming everything edible. They will rest during the heat of the day and do another hunt in the late afternoon, visiting a number of string snares they have set up along various animal trails.

The years in a boy's life spent at the middle fire are both dangerous and traumatic. They are forced to leave the only fire they have ever known when they are driven out by the females before they can present a challenge to the female dominance structure. This expulsion is both bewildering and painful. At the female fire, these boys have some stature, some standing. They dominate the younger children, both male and female. To go from a position of relative influence to a post of lowest rank is difficult. Solace lies in the fact that all the boys must go through this ordeal.

Survival, hard enough at the female fire, is now much more difficult. The boys have some of the skills required to survive, having practiced casting both stones and spears since they were weaned. But now they must give up female guidance and direction to strike out on their own. Some are better equipped for the task than others. Defiant, rebellious young males make the transition with fewer problems than less assertive males. But the change is dangerous and hard for all, and many, almost a quarter in fact, will not succeed.

Males are only grudgingly accepted into the middle fire. At the female fire and at female foraging, each individual finds and eats their own food, wolfing it down before some other more dominant individual takes it away. Even the meat the females take from the males is not shared, the takers getting theirs first, and the rest of the females and children in order. But now this changes. At first, the younger boys at the middle fire are prone to behave as they did at the female fire. They seek out insects, small reptiles, shoots, tubers, and bulbs—anything to fill the belly. But without the protection of the females and the female fire, it is almost impossible. This is why the youngest boys at the middle fire start carrying fire clams. They, more than anyone else, depend on fire for protection.

At the middle fire, the boys must adopt an entirely different behavior pattern. For the first time in their lives, they must begin to think and act in terms of "us" rather than "me." The first hesitant step toward cooperation is taken, and the idea of group loyalty begins to germinate. A young boy, alone on the veld of Africa, is dead. A group of boys are well on their way to becoming what our ancestors were, the top predators of the plains. So boys forget the security and solace of the female fire and begin to bond with middle fire brothers and cousins. They form strong bonds, critical bonds, essential bonds because just like the male fire, each boy needs to know who can do what and how well. Lives depend upon it.

The change is traumatic, and many do not make it. In their rebellion and their individuality and in their independence, they die. Going from the female fire to the middle fire causes almost as many male deaths as the first two years of living on the veld.

Boys that survive, boys that were at the top of the child pecking order at the female fire must now go to the bottom of the middle fire order. Over time, they will work their way up through the order at the middle fire until they are accepted into the male fire, or they die. During the maturing time at the middle fire, the boys will form strong communal bonds with the older, more experienced boys. They will form clicks that owe loyalty only to the middle fire group. In short, they will become part of the gang.

Older boys graduate out of the middle fire in several ways. Some of them are picked by older men who may have recently lost their partner or are establishing a threesome. Other times, two older boys developing their first partnership bonds may come into the male fire as a pair. These boys may continue to hunt with their middle fire gang, or they may trail the males as they hunt, watching and learning adult hunting techniques. They may even attempt to practice these same techniques with the boys, thus smoothing the transition from boys to men.

We lived this way for hundreds of thousands of years. Females, secure at the female fire, spent their entire lives with individuals they have known since birth. For them, there was little of the trauma associated with the transition to the middle fire. Boys, however, faced a different future going through the ups and downs of two fires before they can once again feel a small degree of comfort and security. Because boys will be boys, we should not be too surprised that at a certain age they are prone to form gangs.

Now comes the part that the elite won't like. We have gangs today because young boys find the attraction of the company of other young boys preferable to a life with mother and sisters. It is my contention that this behavior is natural. It is normal, and we cannot expect these individuals to behave any differently for the foreseeable future.

The behavior of these individuals is normal too. The taking of what they want with no concern for others is normal. This is the way they have always behaved. It is also the reason in our past that they were ejected from the female fire. Fighting is normal; killing is normal. All are survival behaviors, all selected for over the millennia.

We will not be able to rehabilitate these individuals because in their minds, they are doing nothing wrong. On the contrary, they are giving vent to deep-seated genetic urges. They are doing exactly the right things, exactly the proper response, just what is needed, and it feels good too.

Do we have gangs today because of the trauma our young male ancestors went through hundreds of thousands of years ago? There is certainly a predisposition, especially among Hispanics and Blacks, to form gangs. The predisposition also occurs among Asians. It is, however, not as well defined in Caucasian cultures. Gang males quickly and completely transfer any familial loyalty to the gang. The gang provides structure and order to the boys' lives. Rules are established, dos and don'ts fixed, and a hierarchy is in place as well as a blueprint for achieving respect and dominance within the gang.

Most gangs are not what we would call today principled. But that, too, has its explanation in our past. To our ancestors, taking was, as I said, uncommon but ordinary. The strong took from the less strong, and in our past, no one was weak. Taking, especially from adjacent groups, was expected and even encouraged. Our ancestors did not have the Stone Age equivalent of pushcart peddlers nor any access to traders. The most likely distribution method was that of the raid and the taking. So a particularly good chocolate-colored obsidian hand axe, over its long lifetime, would have passed through many hands, mostly by violence.

As I have said, we have Black gangs, Hispanic gangs, and Asian gangs, but not so many Caucasian gangs, although they do occur. Any theory of "gangness" must account for this. Blacks, especially those in the western hemisphere, are almost without exception the descendants of slaves. These slaves came from hunter-gatherer tribes of the African interior. These individuals have been separated from their ancestral roots for four hundred years or less.

In Africa, these tribes lived very much the same lives as those individuals who populated Africa 150,000 years ago. Among these tribes, only the strong survived, only the best prevailed, and only the most capable endured. The Blacks of Africa did not have anything in the shape or form of surplus population, and genetic predisposition was very narrowly focused.

It is surplus population that allows for behavioral patterns to exist and multiply outside the restricted hunter-gatherer pattern. It takes many generations and many years for the basic hunter-gatherer genetic predisposition to be diluted to a point where such behavior patterns no longer impact a majority of individuals within the group. Blacks have not yet been exposed to sufficient surplus population potential to substantially alter normal genetic behavior patterns in a number of areas.

But what about Hispanics? Surely, Hispanics have been around long enough to substantially impact normal hunter-gatherer behavior patterns. Hispanics are of European stock, civilized for hundreds of years and diluted with substantial amounts of surplus population. Well, yes and no.

Hispanic male behavior differed substantially from Northern European male behavior with respect to the indigenous inhabitants of the western hemisphere. Spanish males willingly, eagerly, and indiscriminately coupled with native females to a point where Spanish blood was significantly impacted by Native American hunter-gatherer genetic behavioral predispositions. Northern European mixing with Native American females, however, was severely limited by cultural prohibitions, thus reducing the impact of hunter-gatherer behavioral predispositions. This is why there are far fewer Caucasian gangs than there are Black, Hispanic, or Asian gangs.

The predisposition of young males is to seek out and bond with their peer groups. It is stronger in some races than in others; and it is normal, it is natural, and, at one time, it had survival value. We will not be able to eliminate this predisposition. We will only be able to channel and direct this type of behavior.

It is not all bad. Good sports teams are characterized by strong emotional bonds tying the team together. As long as the team can train and contest as a team, it need not have huge amounts of talent. A group with a strong emotional relationship can easily overcome a team having one or two superstars that think and behave as if the sport revolved around them.

Peer-bonding predisposition also shows up in the military. A squad is only as good as its ability to merge the individuals within into a completely integrated unit. Squads that know and understand the strengths and weaknesses of others within the group, who are willing to submerge egos and act as a whole, will be much more likely to succeed. These relationships are emotional as well as mental. There are strong feelings of affection and even devotion generated in a truly successful squad. And it all goes back to that distant time in our ancestral past when to be alone was to be dead.

So far, this book has been an amalgam of fact, inference, and concepts based on what we see in the world around us. It is a fact that we, as intellectual individuals, fail utterly in our ability to view the human animal dispassionately. It is inference that animal behavior, in many ways, reflects the behavior of our ancestors. Lions are predators, and lions do not share. Chimps are social animals. They form dominance hierarchies, are promiscuous, and go to war against other troops. Nature is not nice, and we were, for eons, an integral part of nature. To somehow contend that we were neither touched by nor influenced by nature is delusion.

The history of man has been a history of war, dominance, and a totally lopsided wealth distribution for as far back as recorded history goes. Small highly organized, dedicated elites dominating large unwashed masses. Kings, dictators, monarchs, caesars, the list goes on and on. Dominance writ large, and we, in our delusion, fail to see the pattern. We are nice, we are kind, and we are gentle, and the true nature of human nature is nurture. If only we are all nice to each other, then everything will be fine.

And the truth of the matter is that for most of us, this is true. We do respond to kindness with kindness, generosity with generosity, love with love. Our delusions are just that much harder to overcome because love works for some of us, even for most of us. So why have we had wars, dictators, and death on such a massive scale for such a long time? The accompanying graph does much to explain why our history is the way it is. Understanding the true nature of our alphas will also provide us with a blueprint for better times.

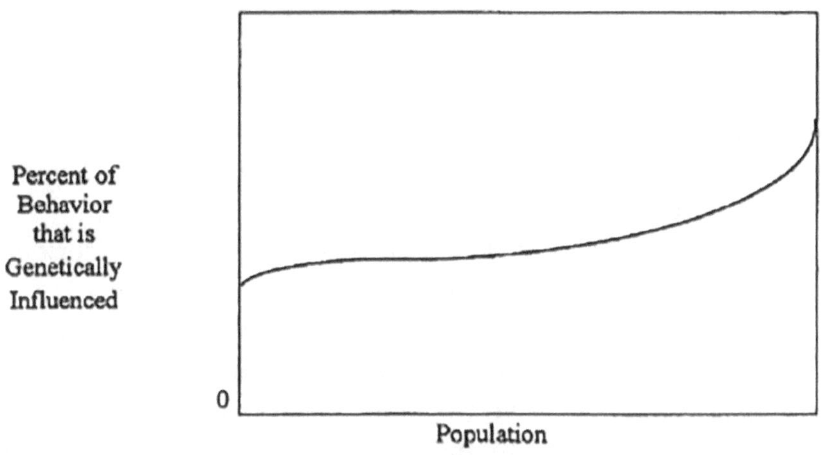

Figure 10-1. Graph of Genetic Influence on Human Behavior

As the graph shows, behavior varies from person to person, but all degrees of genetically influenced human behavior can be plotted on a line leading from left to right. At the left end, we have those who, by an extreme effort of will, attempt to control their biological urges and actually contravene the dictates of nature. Among such people are those who eschew reproductive potential. Nuns, priests, monks, and others who consciously resist the second most import-ant urge in nature fall into this category.

There is no vertical scale, and I have absolutely no idea what it would be. To some extent, it depends on what is included in the definition of human behavior. Breathing is instinctive, that is, genetic, but we can hold our breath and, to some extent, consciously influence our breathing. Should breathing be part of our genetically influenced behavior package or not?

What is important here is the fact that each individual has a unique genetically influenced behavioral pattern. This, coupled with the cultural parameters involved, and individual personality will determine how that individual will behave.

For most of us, 85-plus percent, much of our genetic predispo-sition is inhibited by conscious control and cultural rules. We have urges, we have needs, and we have desires and dreams, but our cul-tures, to a lesser or greater extent, channel and direct these genetic predispositions. The predispositions are also not as pronounced as they were at one time in our past. They are muted and diluted by the force of surplus population.

It is the right end of the graph that is the key to understand-ing human behavior. Some 8 to 10 percent of our population can be considered to be strongly genetically predisposed. These are the alphas; the dominants and their behavioral patterns range all over the map. It is here that we will find dictators past and present—individuals whose thirst for power and control leads to absolute dictatorships and death on a massive scale. Here we will also find capitalists, individuals who lust after power and control but can be forced to achieve their goals within the confines of the culture. In capitalist democracies, they can also be made to supply the public with quality goods and services at reasonable prices.

Criminals are also part of this upward curve. Their acts are crimes only because we say they are. At one time, killing and taking were not only common, but they were also necessary. We were at one time animals, and these individuals still are.

Last but not least, politicians also fall into this area of the curve. Power is the key, dominance the goal, and politicians hunger for the power that election brings. They do not serve to satisfy us, but to savor the sweetness of power, of authority, of importance. This is probably as good an argument for term limits as one could ask for.

It is that small segment of our population on the right that we must come to know and understand—probably less than 10 percent of our population—but the impact of their acts far outweigh their actual numbers. We need to understand that what they do and how they behave is strongly influenced by genetic predisposition. We need to understand that this genetically influenced behavior cannot, for the most part, be stopped, but that it can be controlled. It can be channeled to a point where the alphas—the dominants—can be made to achieve their goals within cultural bounds. How that happens depends on us.

CHAPTER 11

We have always had leaders and followers. It is a character-istic of humanity that we must somehow in some way introduce structure into our lives. We long for leaders, we worship kings, we follow warlords, and we adore the likes of JFK, Gandhi, and Lincoln. These are the leaders, the alphas, the domi-nants. And their position in life reflects their genetic predisposition. These individuals—some good, some bad—have had a profound influence on mankind. It is the behavior of these individuals that we must come to understand if there is to be any possibility of global peace, prosperity, and happiness.

Dominance is the key. Dominance is old beyond man, old beyond mammals millions and hundreds of millions of years old. Dominance is of value to females, as we have seen, because females use dominance to ensure that their offspring have the optimum potential to survive (because dominant males are stronger) and reproduce (because dominant males sire more offspring). But a pre-disposition to dominate had to predate its use as a sexual attractant. Females can ensure that dominance remains a part of genetic predis-position, but they cannot breed it into being. Dominance is deeper than sex; it is primordial, writ large, on the most basic of genetic building blocks.

How deep is dominance? It is a fact that every culture that exists in the world today or has ever existed in the past is simply a dominance system. When analyzed objectively, each culture displays all the characteristics of dominance. Each culture, no matter what its actual beliefs are, makes the following unannounced statement.

If, as a member of this social group, you do what the group considers to be right and proper and important, you will be looked up to, admired, respected, and given status. The more and the better you hew to the group's beliefs and standards, the greater status you will achieve. If, on the other hand, you disregard tribal beliefs, you ignore or denigrate customs and mores, then you will be reviled, despised, and even dispatched. It is always the same: each culture has its own customs, its own beliefs, and its own mores, all of which the culture considers to be right and proper and sound and even reasonable. Individuals growing up in a culture are taught from birth the group's ways. Certain things are done; certain things are just not done. And conduct is judged, and approbation is given or denied based on what the culture considers right and proper.

Man today is more or less civilized. Most individuals in the world either live in civilized environments or have had so much contact with civilization that they would be hard-pressed to exist as their forefathers did. Something as simple as a steel axe or an iron knife or a metal pot—common for most of us—confer tremendous advantage to groups who have only stone axes, flint knives, and clay pots. Once metal has replaced stone, there is no turning back, and even the most remote hidden Stone Age culture knows that.

Simple hunting-gathering is for the most part gone and will never return. The moral arguments of this condition I will leave to the philosophers, but the reality is that once a human hand touches steel, lives are irrevocably changed. That is fact and the basis for all that follows.

There are three main characteristics that define the human animal. We are bipedal, we depend on intelligence for survival, and we have the ability to live in numbers far beyond what nature would dictate for our species. Our erect stature was a direct response to our aquatic lifestyle. We have seen that a change in our cranium occurred in response to that conversion from quadrupedal to bipedal stance. This change eliminated our only natural weapon—our muzzle and large canine teeth; our brain became our last and least hope. Fortunately, it worked. We survived and multiplied, and we did it by using brainpower. No other species has ever done that. And so bipedalism and intelligence define man.

But it is civilization that is the true measure of man. We populate the world in densities no other large mammal even remotely approaches. And it is because of these densities that human nature is so difficult to understand. Civilization is many things, but the defining concept of civilization is possessions.

The Stone Age native in the highlands of New Guinea has an axe. Made of stone, it is smooth and polished, well hafted, and suitable for use in cutting down trees to promote slash-and-burn technology. But his fellow tribesmen also sport axes—smooth, polished, and well hafted. One individual may possess a bow complete with arrows or a spear or a club, but all his brothers and cousins also have these same tools. Thus, this is mostly an egalitarian society. All the individuals within the group have pretty much the same stuff. There may be a bit more ostentation in a chief's house or clothes or decorations or tools, but a stone axe cuts down trees, whether it is pretty or plain. A bow and arrow is useful regardless of its looks. Reality requires few possessions, little or no variation in tools or living conditions, and simple hierarchal social patterns.

Possessions in quantity changed all that. It started with grain storage. Unlike meat, grains can be stored to be used at some future time for either sustenance or seed stock. It started with the taming of goats to be used as a source of milk and meat. It then advanced to the possession of land, land that could grow grain and pasture goats. It then progressed to ownership of the rude stone huts next to the fields where grain was grown and herds were kept. Some fields of grain were better than others. This made them more desirable to individuals or to groups. Some herds were larger than others, had more and better animals, and were less wild and more productive. Some dwellings were better than others—closer to the springs, in the shade of the all-too-few trees, even better built. And so, over time, we gradually modified our dominance systems. As we moved from hunter-gatherers to farmer-herders, the rules changed. The alpha was no longer the biggest, meanest son of a bitch in the valley. That position came to be occupied by the individual with the largest herd or the best field or the biggest house in the valley. Dominance persisted, but the measurement of dominance changed. Then something really strange happened.

A man found out that he was as much responsible for the birth of a child as the woman who bore that child. It was as much his as it was hers. We have been forced to live with the results of this knowledge ever since.

I am sure that the female of the species knew for a long time that a man was necessary for reproductive success. After all, the same orifice was used for both sex and birth, and no woman in any woman's memory had ever given birth without first having a sexual experience. The woman may not have known the exact relationship, but they were aware of the broad outlines, and one day, one of them blabbed. It is indeed possible that this occurred more than once.

Children became possessions, part of a male's worth, and because the females were the ones who bore the children, they, too, had to become possessions. After all, the male had to ensure that the offspring produced were his and his alone. He had possessions to pass along. Thus, a naturally promiscuous animal attempted to enforce monogamy on the female while the male remained promiscuous. As we have seen, we are not monogamous animals. Now we see why the male of the species has tried to force monogamy on the female while the male remains polygamous.

Possessions and a mythical monogamy define our present lifestyles. Much of the good and the bad of our current condition springs from these two concepts.

Civilization is both a boon and a bane. We live longer, healthier, less dangerous lives under civilization. There is the rule of law, order, and structure to civilized society. In short, no matter what some individuals say, it is better to be civilized.

But civilization is not all tea cakes and roses. Only very few civilized societies achieve the kind of success that truly gives the greatest potential to the greatest number of individuals. And there are still the dominants that must be considered—the alphas, the leaders, those genetically inclined to be on top. Civilization has bred much of the predisposition toward "alphaness" and dominance out of us, but in some, it still throbs. How this is handled goes a long way toward determining the kind of civilization we achieve.

Let us then recapitulate. We are evolved animals. For hundreds of millions of years, our ancestors relied on instinct and genetic predisposition to survive and reproduce. In this matter, we succeeded, else I wouldn't be here to write this, and you would not be here to read this.

It has only been in the last two or three million years that, in our species at least, survival and procreation has relied on anything other than instinct and genetic predisposition. And it has only been in the last two hundred thousand years that intellect has had any appreciable impact on our behavior. It has only been during the last forty thousand years that we have, to any degree, stepped beyond our instinctive bonds. Until ten thousand years ago, certain very specific, very simple, and very self-serving behaviors were the norm. We were mean because mean survived just as it does today in most animals. We were vicious for the very simple reason that killers always live longer than their victims. We thought of our own needs and wants first, last, and always, but with one exception. Our survival depended on cooperation within limits. We were loyal and devoted to our partner or partners and supportive of the clan against all other clans. But by and large, we were just as mean, dirty, nasty, and animalistic as any other species in the world. We were that way because that was the way of survival.

There was no justice, no mercy, no sharing, no caring. Hostility reigned, xenophobia prevailed, and we survived. Certain behavior patterns were successful. Dominance was chief among these behavior patterns. Individuals who behaved in a dominant fashion left more offspring than those who did not compete. It is as simple as that. Differences of only 2 or 3 percent over time will result in certain characteristics becoming paramount. We all lived the same way. We existed in smallish bands—competing with our neighbors, defending territories, establishing hierarchies, and surviving and procreating.

For two hundred thousand years, certain very specific, very simple behavior patterns got us through. Then ten thousand years ago, things changed. The change occurred in fits and starts. We slowly developed the realization that these certain very simple, very specific behavior patterns were not the only way to survive and procreate. Individuals who would not have survived in hunter-gatherer groups could in farmer-herder groups. And so civilization was born, and with it surplus population.

Over time, social groups diluted but did not erase all the very simple, very specific behavior patterns of our hunter-gatherer ancestors. As a result, we see a range of behavior patterns in different cultures in the world today. In cultures that have had surplus populations for longer periods of time, we have fewer individuals who exhibit hunter-gatherer dominance-oriented behavior patterns. In cultures that have little or no surplus population, we tend to see behavior that more closely mirrors that of our ancestors.

The real key is to know and understand why we behave the way we do. We are—each one of us—influenced to a greater or lesser degree by our evolutionary past. We all behave, at times, in ways that can be traced back to a genetic predisposition, a set of behavioral patterns that at one time offered survival.

Webster's Dictionary defines culture as "the concepts, habits, skills, arts, instruments, and institutions of a given people in a given period." Cultures also include beliefs, laws—both written and unwritten—dress, division of labor, and hierarchies. And all cultures share the same basic premise. If you dress right, believe right, act right, and exhibit patience, you will achieve status within the culture. Cultures are, one and all, dominance systems.

Consider the Wayana, a small tribe of South American Indians living along the Maroni and the Itany rivers in French Guiana. Numbering about five hundred souls, the tribe makes a meager living from fishing, hunting, and slash-and-burn farming. There is a tribal hierarchy with chiefs and, well, Indians. There are certain rules and regulations and rites that are followed within the tribe. Among these is the rite of pain called *marake*.

The procedure calls for the initiate to withstand the bites of hundreds of black ants that have been woven into a mat frame. The frame is placed against various parts of the initiate's body. The bite of the ant is very painful, but it is considered bad form to cry out in agony. The rationale is that life is hard and painful; you might as well get used to it. Successfully completing the ordeal does enhance one's status.

The Wayana have a set of rules and regulations, ways and beliefs—many of little or no concrete value but observed nonetheless. Those who obey and observe are looked up to, admired, and given status within the culture. Any Wayana that transgresses the belief system of the tribe is shunned or even expelled. Wayana culture is a dominance-determinative system.

Or consider the Zulu. They were one of the largest single population groups in South Africa. At one time, Zulu culture was based on hereditary kingship with a supreme king and numerous divisional kings. Life among the Zulu was regulated by a complex code of conduct for both sexes. Men had specific duties such as building huts and herding goats and cattle. In former times, a man's responsibilities also included military service to his king. It was only after his term of enrollment in the military was completed that he could seek out a wife and start a family. Each man, during the course of his life, would pass through a series of well-marked steps, each of which required certain preconditions, preparation, and ceremony. Each step also increased the status and importance of the individual.

Zulu women also were expected to go through a series of preparations and ceremonies to reflect and demonstrate their relative position within the culture. It was always the same—do this, do that, and you will be looked up to and admired and respected. Failure to follow procedure would bring shame, reprimand, and sometimes death. Dominance is that important.

Another example of the culture of dominance is that of the Pentecosters, a group of islanders who live on Pentecost Island in the nation of Vanuatu. Pentecosters are land divers, the original bungee jumpers but without the rubber ropes. These people (males only) tie vines to their ankles and jump from rickety spires of wood some seventy-five to one hundred feet high. The vines are supposed to be just long enough and the towers just resilient enough to allow the jumper to barely touch the ground before being stopped by the vines. They have been doing this for centuries, and by and large, it works. Within the memory of the land divers, only one man has died as a result of land diving.

The diving takes places in the spring and is supposed to ensure a good yam crop. The higher the jump, the higher the yam foliage and the better the crop. But reality is that the higher the tower from which one jumps, the higher the status. High jumpers are looked up to, given status, and admired. Land diving in the Pentecost Islands is a dominance thing, with those jumping from the highest towers obtaining the highest status.

This cultural predisposition with status and dominance is not limited to today's cultures. The structure of culture to reflect a dominance hierarchy has existed as long as cultures have existed. In the beginning, it was hard enough just to survive without loading down the lifestyle with a lot of dos and don'ts to muddy up the waters. Besides, within the social group of no more than six to ten males, each member knew his status, and so status-reflecting behavior consisted of challenges and backing down, or not. In any event, everyone knew who was boss.

It was only when our ancestors amalgamated into larger groups that some method was needed to distinguish alpha from beta from gamma. This was also the time when the herding-farming lifestyle made it possible for these people to develop a bit of sustenance security. This opened up time to come up with various ceremonials and rituals to determine who was who. Simple cultures, simple rules; complex cultures in our past or in our present, complex rules and regulations and beliefs and mores and behaviors.

Consider the Spartans. These were the big boys when it came to ancient Greece, some of the best warriors in the world, and they should be. A male child was taken from his mother at age seven. For the next twenty-three years, his life was one of preparation for and the practice of war. Certain procedures were followed. The child was kept barefoot to ensure tough, hardened soles as a part of a tough, hardened body. Children were poorly fed but encouraged to steal food but not get caught. Such behavior was looked up to, admired, and a Spartan capable of and willing to do anything for Spartan society was a respected Spartan.

All this training and experience was for a single purpose. A good Spartan male was a good warrior. War was the be-all and end-all of Spartan male society. To be tough, resourceful, cunning, and skillful with weapons was the epitome of Spartan life. And so Spartans trained all their lives to live up to Spartan standards.

And Spartan standards were spartan. Emphasis was placed on simplicity rather than opulence, homespun rather than imported, and plain rather than luxurious surroundings. A fetish was made of plain gruel—as food, a cloak, and a bed of earth for sleeping—and a singular lack of female companionship.

All this in spite of the fact that the Spartans held undisputed rule over the Greek city-state and could have had anything they wanted. Spartan culture was built on war and warriors, on being prepared for war, on being able to fight and win. Those who hewed closest to Spartan standards were accorded the highest status. Spartan culture was a dominance system.

Not all dominance behavior is bad. This predisposition can be used by the masses. In Papua New Guinea, there is a tribe called the Kawelka. To achieve status and dominance among the Kawelka, one must hold a *Moka*, better known as a big-man ceremony. The name says it all. In order to be a big man, a Kawelka male would go about and exhort his kin group to put together food and presents for a big feast. His skill at organizing and coordinating this activity will directly affect his standing in Kawelka society. A good feast with lots of food and plenty of presents will ensure high status within the culture. In fact, one of the goals of the big-man ceremonies is to outdo the last big-man ceremony.

And big men brag. The real purpose of the ceremony is not to feed the participants. Its only purpose is to allow the big man to brag on how wonderful, how gorgeous, how spectacular a party he threw. This is especially true if he has invited another big man to the feast. One-upmanship is old.

The Kawelka of Papua New Guinea are not alone in handing out presents. The Haida of northwestern North America had a similar custom called potlatch. Years could be spent getting enough wealth together to hold a potlatch, where everything would be given away in a show of power and strength.

These ceremonies, while they benefited all the people of the tribe, were not designed nor instituted for that purpose. The primary purpose of both big-man and potlatch ceremonies was dominance. By pulling off a great ceremony, one achieved status within the social group. If these individuals had been asked to do the same thing while remaining anonymous, they would have refused. The ceremonies were not the result of altruistic feelings or a desire to do good. Both big man and potlatch ceremonies are aspects of a dominance hierarchy within the culture. They are used not to do good, but to dominate.

These examples of what can be described as cultural customs reflect a belief pattern unique to each cultural group. I have used the giving custom of the Kawelka and the Haida for a specific reason.

These cultures are, at times, used as examples of the basic goodness of man. See, we are told, these individuals are doing good for the group. This reflects the basic nature of man, which is giving.

Reality intrudes. Potlatch ceremonies and big-man feasts do reflect the true nature of man, and it is not giving; it is status. A big man acquires status by holding feasts; he acquires the wealth of the culture because he is looked up to, admired, and given status. He acquires that which he desires—the admiration of the culture. This wealth, this coin of the realm, this status is what the individual is really after.

We will get into cultural cash shortly, but first, a short digression demonstrating the fact that we want so desperately to think well of ourselves that we cannot think clearly.

There is a verbal and quite vocal squabble among the paleo-anthropologists over the descent of modern man. There is the out-of-Africa school that says we are descendants of one female named, appropriately enough, Eve who lived in Africa some two hundred thousand years ago. The second school says, no, we gradually changed from *Home erectus* to *Homo sapiens* over a period of time all over the world. Individual populations would evolve in place, each turning into *Homo sapiens* from *Homo erectus* in response to natural selection. This is known as the multiregional theory. It implies that all individuals from the birth of *Homo erectus* two million years ago belong to the same species. Something some anthropologists find hard to believe.

The debate would just be another minor scientific dispute except for the implication of the out-of-Africa theory. Under the multiregional theory, we just quietly and peacefully evolved from *Homo erectus* to *Homo sapiens*—no muss, no fuss, no bother. Out-of-Africa, on the other hand, screams takeover: "*Sapien* hoodlums manhandle *erectus*." And this is the problem. If *Homo sapiens* spread out of Africa and took over what was then the entire known world, the only way they could have done it was by eliminating *Homo erectus*. This is anathema to most multiregionalists because it impugns the noble character of man. We are kind, we are nice, we are noble, and we would not—under any circumstances, especially in a setting so benign as nature—be so beastly. But we are here, and *Homo erectus* is not, so something happened.

The concept of a noble, peaceful, caring, human ancestor is a delusion. It works nowhere in nature, and we are deluded if we think we were the lone exception. Mean works; nice doesn't. We see it so often in nature that it almost becomes blasé.

Lions are a perfect example of nature. Lions live in prides, which are female centered. Males come and go, but grandmothers, mothers, and daughters form the core of the pride. Male offspring, as they become lion teenagers, are driven out of the pride. They must fend for themselves or die. When they are old enough and strong enough, they will try to take over a pride.

And here is the rub. When a new male or males take over a pride, they kill all the cubs. It is mean, it is nasty, but it is nature, and it works. By killing all the cubs sired by the previous pride male, the new male brings all the females back into a state of estrus. The new male then impregnates the females. It will be his offspring and only his offspring that the pride will produce for the two or three or four years he is in charge. Then a new male will take over, kill all the cubs, and begin again.

This is the reality of nature. Evolution rewards takers; as a result, most of what we see in nature is takers. Takers can be as small as a flea—a bug that sucks your blood—and, in the process, can infect you with any number of deadly diseases. That doesn't seem to matter to the flea.

We see man's cruelty to man as an aberration when, in fact, for hundreds of thousands of years, it was de rigueur. Kill or be killed; fight or die, and it worked, for only the strongest, most able individuals were able to survive. Anytime an individual gave up his or her life for another, it immediately stopped any chance of such behavior being genetically passed on. Only under certain very restricted, very personal circumstances was self-sacrifice selected for. And that was for bond mates.

But we want us to be nice, to be kind, to be altruistic and caring. There is a bias among anthropologists toward making us a fitting cap to evolution, a nice species. The bias shows up time and again in our literature, and no example is more significant than the present-day interpretation of the footprints at Latoli.

The main sets of prints consist of the tracks of two individuals traveling in the same direction. Illustrations of the individuals making the prints show a larger individual, maybe male, with his arm around the shoulder of the other individual, presumably female. Thus, we show the kindness and caring of our prehuman ancestors. Thus, we preserve our delusions.

The prints were, in all probability, not made at the same time. According to Mary Leakey, the prints are so close together as to rule out the probability of a pair traveling abreast. Another clue comes from the prints themselves. The smaller set is clear and crisp while the larger is blurred, indicating that the smaller set was made while the ash was damp. The larger set, on the other hand, was probably made either before the rain or after the ash had started to dry up.

Still we dream, still we hope, and still we ache for the nicety of our human ancestors and even their ancestors. This is as good an example of anthropological delusion as could be asked for.

*****

Cultures are also measurement systems. Behavior within the culture is measured and accepted or found wanting. This measurement system can be referred to as the coin of the realm. Different cultures have different coins, but the measurement system is always the same. In every culture ever studied, there has always been found a coin of the realm. Accumulate such coins and achieve status, power, and dominance. Squander such coins and lose your position within the culture and maybe even your life.

The coin of the realm is that which the social group treasures. It is that which the culture holds dear, and for many—indeed, most—social units, it has nothing to do with money. For most cultures, it is behavior that constitutes coin of the realm. Individuals behave in certain ways, and the culture finds that attractive. Desirable behavior on a regular basis results in the accumulation of coin of the realm within the group.

Among the Wayana, as we have seen, to undergo the *marake*, or ordeal of the biting ants, accords the bearer status. Such people are looked up to, admired, and accorded rank within the community. They have accumulated coin of the realm. The more times someone undergoes the rite, the more approbation they get.

But there are other behaviors that also define the Wayana, and one of them is the curious custom of *kasili*. *Kasili* is a mild fermented beverage made from manioc root. It is considered good form to drink *kasili* to capacity and then vomit the imbibed beverage. Custom requires that the imbiber then consume more *kasili* and vomit yet again. In this way, the small amount of alcohol in the brew is multiplied, and individuals can become mildly intoxicated. Such behavior is considered by the Wayana to be good table manners.

Then there is the downside to Wayana life. Unlike *kasili*, which has little alcohol, rum provided by the traders makes the Wayana stumbling drunk, and among the people, this is considered very bad form.

Among the cultures with little material wealth, it is behavior that is the basis for acquiring status within the group. The Wayana have little in the way of wealth. They are subsistence farmers using slash-and-burn on communal grounds along with fishing and gathering as a means of sustenance. Until first contact with the Europeans, they were essentially Stone Age in living patterns. They had no pottery, no draft animals, and no metal. Their development of relative position within the culture was based on kingship, kinship, and behavior.

Among the Wayana, there is a leader and a quorum of senior men who had influence and power within the group. But if the followers did not like a leader or thought his decisions were bad, they could overthrow him and indicate a preference for another. It all depended on behavior. Do the right thing and prosper; do the wrong thing and face failure.

Among the Kawelka and the Haida, it is a combination of behavior and wealth. Months, even years, are spent accumulating enough wealth so that it can all be given away and so the giver can brag up his giving.

Among the Spartans, who had all the wealth of the Greek city-states of Sparta at their disposal, it was spartan behavior that brought status and dominance to an individual. It was only after a male Spartan had turned sixty that he could begin to partake of the real wealth of Sparta.

But what happens when behavior patterns that have been in place for thousands of years no longer determine status? What happens when the coin of the realm is no longer spendable? We need look no further than the Indians of North America for an answer. The North American Indians provide us with a perfect example of what happens when Stone Age cultures come into contact with wealth and power. The American Indian story is one of tragedy and death, but it tells us what happens when a culture is stripped of its status determinatives. As we have seen, the Indians were newcomers to the Americas. Africa is the cradle of mankind, and our African history goes back hundreds of thousands of years. Europe and Asia have been populated for tens of thousands of years. But the Americas have had human populations for only the last fourteen to sixteen thousand years.

Territory, at first, was easy. There was lots of land, and the natives spread out in great waves, taking advantage of both flora and fauna. But soon, all the territory was occupied. Now things got a little tighter. Animals unafraid of man were dead. Animals afraid of man were difficult to hunt. For a small group of hunter-gatherers, a deer or two a week would work. But as social units grew, game in the immediate vicinity became scarce. Now it was necessary to range farther afield to find sustenance. Eventually, a social group came in contact with other social groups. Weapons of the hunt could also be weapons of war and were so used.

Boundaries were established, territory was occupied, but the borders were iffy. War became a way to maintain sufficient room to live and breathe, as well as a way to acquire women and slaves. Over time, a culture of war was established among the Indians of North and South America. Waging war became the coin of the realm for most Indian tribes, especially of the plain and prairie Indians of North America.

Plains Indians had little in the way of material wealth. They had no draft animals, except for dogs, which could carry little. This meant that whenever the group moved, it had to pack all its belongings on its collective back. This was strong inducement to keep possessions small and simple. One could not acquire social status by acquiring possessions. In a culture where additional possessions were liabilities rather than assets, behavior became the coin of the realm.

Among males, acquiring coin of the realm followed two entwined paths—hunting and war. A good hunter was looked up to and esteemed because he could take care of his family. At the same time, the skills that made a good hunter also made a good warrior. The ability to sneak up on a deer translated easily into the ability to sneak up on an enemy camp. Hunting weapons were also warfare weapons—bows and arrows, spears, knives, and clubs. A hunter needed to be strong. A warrior also needed to be strong. Good eyesight was necessary for both hunting and war. Hunting was important. While some gathering took place and even farming in spots, hunting was a primary source of foodstuff. A good hunter would have a reputation as such and achieve status as a result, but not anywhere near the status accorded to warriors.

Reputation, status, and influence among the Indians depended upon being a successful, daring warrior. War was also a matter of honor. To be brave, to be bold, to be fearless—this was the essence of warrior behavior; this was the warrior coin of the realm.

Killing, while it was often the result of raiding party activity, was not the chief goal. A real warrior, a courageous warrior, an aggressive warrior wanted to count coup. To touch the enemy—that was the goal. With the hand preferably, but a lesser honor if the bow was used or the coupstick, a device deliberately fashioned to perform such a deed. The enemy, too, was part of the equation. Counting coup on an unwounded, armed enemy confirmed a higher honor than touching an unarmed or wounded or dead warrior.

But what good is bravery in battle if no one knows? The Indians had an answer for that too. There were two ways in which a warrior could display his acts of valor. A warrior used special insignia to designate his individual exploits. Perhaps the right to paint his face in a certain way or maybe an eagle feather or a special article of clothing—items of little or no intrinsic value but of priceless significance to the owner.

But the real way a warrior demonstrated his status within the social group was brag about it. Indians admired oratory. To pontificate eloquently was the epitome of warrior status. To recite in minute detail the exploits of the war party and the bravery of the individual doing the talking was the height of status. To have the whole band hanging on your every word as you acted out the counting of a coup or the theft of a horse—this was the goal of the Indian warrior.

How important was the war party and the deeds connected with it? Among the Crow, a tribe occupying the high plains of North America, a warrior could not become a leader or chief until he had counted a coup, stolen an enemy bow, taken a tethered horse from an enemy camp, and led a war party.

War was the basis for status and dominance within the tribe. A warrior may only have one bow, one knife, and one spear, but he could become rich in deeds. When the White Man came, he took away the Indians' coin of the realm. The White Man expected the Indian to adopt the White Man's coin of the realm, or to just go off quietly and die. A few did, indeed, adapt to the White culture of wealth equals status, and many more died. But the vast majority of Indians were left with no way to formalize and structure their cultural values. It should not come as a surprise that the Indians of today live in poverty and alcoholism. We have taken away their reason for living, and neither the Indians nor we have anything to replace that.

The Indians fought among themselves for several reasons. As we have seen, warrior status was a prime reason for war, but there were other advantages to waging war as opposed to living in peace. War kept population densities down. War provided the tribe with females and slaves. Foreign females, although the Indians didn't know it, improved the gene pool.

Indian reasons for war were simple and straightforward. Status, honor, and bragging rights all accrued to the successful warrior. That was reason enough. But later, larger wars, other wars, need other explanations. Still the answer is the same. As one warrior could dominate the tribal campfire recounting his heroic exploits, another warrior could trace his name in the dust of history by trying to conquer the world. It is still a matter of dominance.

War is nothing more than dominance writ large. It is not countries that go to war but men, men bent on dominance and on dominating others. All such individuals dream of a world under their control, responsive to their whims, and subservient to their demands. This is what drives dictators and starts wars. This is what we must understand if we are to put an end to warfare.

We see war all around us, and we try to explain that tragedy. World War II was caused by the Allies being too hard on Germany after World War I. Japan would have been nice to us if only we had been nice to Japan. So the war in the Pacific was actually the fault of the United States. Never mind the fact that Japan invaded China and Manchuria in the thirties and had had an expansionist agenda for decades. We have to have fixable reasons for war. We have to have logical reasons for war. Because, at heart, we are really nice people and just would not go to war without a good reason.

But we also intuitively understand the dominance factor engendered in warfare. And we also understand, without explanation, the powerful drive of dictatorship. When the *Star Wars* series first appeared, we all understood. There was the need of the emperor to have complete, unquestioned dominance over the entire universe. There was the small band of rebels bent on wrecking the emperor's mad scheme. This state of affairs did not have to be explained to us. No one had to point out or explain the emperor's motivation. It was not necessary to treat the audience to flashbacks of the emperor's childhood and possible traumatic explanations of his behavior. We all know; we all understand. It was dominance, the genetic predisposition to dominate. And everybody understood and accepted without question the emperor's motivation. We all understood the reason for *Star Wars*. It is this same reason that is responsible for most wars.

War has been with us always. When Narmer united upper and lower Egypt five thousand years ago, he did it by war and conquest. Since then, the history of Egypt has been a history or war. The first battle that we know anything about in any detail was fought between Ramesses II and the Hittite king, Muwatallis. Ramesses claimed victory, but history indicates, at best, a draw.

Egypt was unique in one respect—its location. With deserts on either side, the country was pretty well isolated from danger. The nomadic tribes that lived east and west and south of Egypt posed little problem. Egypt could have lived in isolated, relative luxury if it had not been for its alphas. Time and again, they marched out of the Nile Valley to do battle in the Middle East. All the great Egyptian kings were great warriors. Not one is fabled for the many years of peace he brought to Egypt.

And it wasn't just Egypt either. It actually began in Jericho ten thousand years ago. Archeologists have found a tower thirty feet high and part of a fortification. Ten thousand years ago, we were at war, almost as soon as civilization began.

One of the first alphas of the Middle East was Sargon of Akkad. Around 2400 BCE, he established one of the first empires of note in western Asia. Sargon and his descendants ruled this kingdom with greater or lesser success for a number of generations.

Next we see the rise of Hammurabi of Babylon. Known for his laws, Hammurabi was also a mighty alpha and extended his rule over much of the Middle East.

Then comes the Trojan War. Was it real? Did it actually happen? Maybe yes, maybe no. There was a Troy, present-day Hissarlik, exactly where it is supposed to be, according to the *Iliad*. There was also a Mycenae on the Greek mainland. If there was a Priam or a Helen or a Hector or an Achilles, we may never know. We do know of at least nine different Troys, each built upon the ruins of the previous Troy and each, in turn, sacked and destroyed in some demented war. Troy and its ashes speak of war, of dominance, of power, and of man's attempt, time after time, to subjugate man.

After the trial of Troy come Sargon II and Tiglath-pileser III. They came out of Assyria to build an empire based on war and destruction. The history of the Middle East is a history of war. Ashur-nasir-pal II, another Assyrian, and Shalmaneser III, yet another Assyrian between them, spread their dominion to the shores of the Mediterranean Sea. Nebuchadnezzar, in turn, captured Tyre and Jerusalem. Then there was Cyrus the Great, Xerxes, and Alexander the Great.

Alexander, of whom it is said, "He conquered the world and wept because he had no more world to conquer." Alexander was Greek, and he extended Grecian domain all the way to India. But the real Mediterranean powerhouse was Julius Caesar, who extended Roman domination from the Middle East all the way to England. Caesar's motto was "I came, I saw, I conquered." Julius Caesar was an alpha.

The history of Europe is a history of war such as the War of the Roses, the Hundred Years' War, England against France against Spain against Rome against the Low Countries—a history of war. There was Napoleon and his disastrous Russian campaign. The saga of the Spanish Armada and the Englishman Sir Francis Drake is a story of war. Wars for dominance, wars for power, wars for no other reason than to kill thousands of men and satisfy an alpha's lust for power.

China, too, has a history of war. The Great Wall was built to protect China from warring tribes on the outskirts of the empire. Central and South America under the rule of the Incas and the Aztecs was a rule of dominance and war. We have already seen that war was a primary occupation of North American Indians who built a culture of status and dominance on the practice and implementation of the arts of war.

If there is any doubt that war and dominance are the hallmarks of man, war in paradise should put that doubt to rest. Picture paradise. Cool, balmy temperatures of, say, seventy-five to eighty-five degrees occur year-round. Islands that have beautiful, pristine sand beaches, and fruits and vegetables that spring up from the ground or fall from the trees. There are banana, pineapple, mango, coconut, taro; and the sea, a deep blue green, full of fish and shellfish, a brimming basket waiting to be taken; and pigs, big, fat, and imminently eatable. That's paradise, a place of lazy days, warm nights, cool breezes, and lush vegetation.

One would think that in such circumstances, the true nature of man might manifest itself. An isolated, self-sufficient utopia where there is little or no need to exert one's self or to display feelings of envy. No reason, really, when everyone eats well, sleeps well, and lives well. If we, as a species, are nice, kind, altruistic, and unselfish, surely living in paradise should demonstrate that this is the way we naturally are.

Not so. The history of Hawaii from its first discovery by the Polynesians right up to and during early European exploration has been a history of war. At one time, each of the islands was ruled by its own chief, and in some instances, there were three or four chiefs contesting for territory on one island. It was a story of acquiring additional territory and the defeat of the nearest rival. In the late 1700s, one chief had consolidated his rule over all the islands except the big one. And it was then that a full-fledged war broke out in the islands. King Kamehameha defeated King Kahekili and established his dominance over the entire island chain.

What possible reason could Kamehameha have for doing what he did? What solid, sound, and believable reason can be offered for warfare in paradise? There is none. There is no reason except an inordinate drive toward dominance, a deep genetic need to know and have everyone else know that there is only one alpha, one dominant.

It is not just Hawaii either. The history of Micronesia, Melanesia, Eastern and Western Polynesia has been one of war. Chiefs and warriors sailed thousands of miles to fight some tribe for some stated reason or for no reason at all other than the fact that it has always been done that way. It was not pretty; it was not good, but it was normal, ordinary, and reflective of the true nature of man.

War can be stopped; war can be prevented, but it will continue to occur until we come to understand the root of war. It's all about dominance, it's all about status, and it's all about control. It's about the true nature of man.

*****

To know and understand why we behave as we do is important. Our failure to grasp the true nature of man has, time and time again, caused individuals to make assumptions that turn out to be false. Some were minor and mattered little; some were huge and led directly to the deaths of tens of millions. Examples abound.

We need to look no farther than England and Germany just prior to World War II. The English prime minister, Neville Chamberlain, reached an agreement with Hitler that in return for allowing Germany to occupy parts of Czechoslovakia, the Reich would agree to settle all future concerns between England and Germany through consultation. In other words, Germany and England would solve any problems between them by holding talks. Chamberlain proclaimed, "Peace for our time," urging the English to "go home and get a nice quiet sleep." Eleven months later, Hitler invaded Poland and started World War II.

Chamberlain was deluded. He assumed that Hitler could be reasoned with. He thought Hitler's goals were modest, that Hitler was, at heart, rational and reasonable, someone who had no desire or need to start a large European war.

Hitler was not rational. He was not reasonable nor logical nor understanding. Hitler was an alpha, and the reason he started World War II was dominance. His goal was to dominate Europe, then Asia, then the world. We are lucky he didn't succeed.

Chamberlain did not understand Hitler's motivation. He and England paid dearly for that lack of understanding. The German people failed utterly to understand Hitler. And they, too, paid dearly. Had the free world stood up to Hitler rather than attempted to appease him, it is possible that World War II in Europe might never have occurred. Alphas do weigh risks; they do judge ability and resistance. The fact that no world wars have occurred in the last fifty years indicates that alphas can be held in check, but only if free people everywhere know and understand the driving force of the alphas among us.

It is always the dictators that start the wars. It is the kings, the autocrats, the caesars, the dominants that set the ball rolling. Their questing for power, for possessions, for dominance is what we need to keep in mind.

The question is not "Do we have dictators?" but "Why do we have dictators?" That dictators exist today is indisputable. They appear in myriad forms. Castro, Gaddafi, Hussein, Assad—all are dictators, one absolute system, one line of authority, one man or group of men on top. China, most of the countries of Africa,

Asia, and until fairly recently Central and South America including Mexico were all dictatorships. Russia, that great bear of a country until very recently, was under the thumb of dictators. Khrushchev, Brezhnev, and Stalin right on through to the czars—who, at least, had some small claim to rule—were all dictators. Even Europe, until recently, was all kingdoms, having rulers whose rule was absolute and unlimited.

Cultures large and small, rich and poor—all had one individual at the top, with a single ruler supreme and omnipotent. As far back in recorded history as one cares to go, we see individuals who aspire to dominate and countries blighted by dominants.

Much of our history is the history of dictators of one stripe or another. Much of our current and past trouble can be traced to dictators and religion, which is a form of dictatorship or would like to be. Hitler was a dictator, an absolute ruler, first of Germany and then of most of Europe. His life equals the lives of more than a hundred million people killed, some accidentally but most deliberately. All because one man wanted to rule over a greater and greater number of people. The need to dominate, the need to assert and maintain absolute control over others is in our genes, and it will always be with us. It is important that we understand why. It is even more important that we know how to deal with would-be dictators. They can be controlled; they can be made to conform to cultural requirements, but it is the responsibility of the followers, the dominated, to lay out the rules and regulations permitting their domination.

There is an ongoing low-key debate among the bastions of academia. Should the citizens of a country be held responsible for the wars waged by their leaders? There are those who say that the citizens of Japan were innocent victims of World War II and that the United States should not have dropped the atomic bomb on them. Some kinder, gentler way should have been used to bring World War II in the Pacific to a conclusion.

But think about it, the Japanese military could not have waged war without the active participation of the Japanese people. The citizens of Japan built the ships and planes and tanks and guns used by the Japanese military in their conquest of China and Southeast Asia. They provided the food and medicine and clothing needed to prosecute the war. The Japanese citizenry provided the sons and fathers needed to wage the war. But most importantly, the citizens of Japan provided a culture of approval that made it possible for the military to make war.

The citizens of Japan could have said no. The citizens of Japan should have said no. If they had any doubts at all about the viability of the war, they should have expressed them. They did not, and they reaped the consequences of this failure.

If I am a Japanese citizen making bullets for the army, do I not have any responsibility with respect to the results of my actions? Is it not, in the final analysis, my duty to determine what possible consequences my production might have? Do I not have an obligation to consider what effect making bullets might have on my wife and children, my city, my country? Should I be held responsible for my actions?

There are those who say that the citizens of Japan are not responsible because they did not know, that they were ignorant and gullible dupes used by their military to accomplish their nefarious ends. Ignorance is innocence, and that trumps everything else.

The argument might wash if the Japanese citizenry had no way of knowing what might be the consequences of their actions. But history records instance after instance of what happens to losers. Indeed, the Japanese military demonstrated, in detail, what happened to loser Koreans and loser Chinese when they fell victim to Japanese ambitions. Could not the citizens of Japan, at least, consider the consequences they might face as losers if they fell to Allied wrath?

The Japanese people could have stopped World War II. Some of them, even many of them, might have been killed in such an action. On the other hand, many of them died anyway, and for a lost cause. Followers need to know that they and they alone have both the ability and the responsibility to control the actions of their leaders.

It is time to make another politically incorrect assertion. It is my contention that the individual will always and in all ways be held responsible for any and all actions and inactions the individual chooses to make. It does not "take a village." The responsibility will always and in every way be borne by the individual. It is not as if the individual has a choice. It is not as if laws and rules can mitigate responsibility. In the end, the individual will have to deal with the effect of the individual's action.

A recent editorial in one of the local newspapers demonstrates the collective responsibility delusion of the elite. It also offers a point from which to start our quest for reality in the delusions of men.

The editorial is shocking. African Americans in the United States account for almost half of new AIDS cases even though they make up only 13 percent of the American population. AIDS is the leading cause of death among Black females between the ages of twenty-five and thirty-four. And, we are told, the reason for this condition is silence. We do not talk about AIDS enough. We do not discuss sex enough. We do not acknowledge the fact of drug use enough. On and on it goes. We are responsible. It is our fault this "plague" exists. It is communal; it is not individual.

But think about it. It is not "us" who are going to die of AIDS. "We" are not going to suffer the wasting and debilitating effects of AIDS. It is the individual who has AIDS who is going to suffer.

Perhaps a more open and honest communication would have helped. I don't know. I would, however, be willing to bet that its impact would be minimal. That is not to say that I am opposed to a program of information distribution. It may help, but it is not the answer. It is always the individual who pays, always. We may not like this. We may not want this. We may wish to think that this is an outrageous, infamous, crying shame. No matter, the individual will always be the one who is responsible for individual behavior.

That being the case, perhaps it is time to make people responsible for their actions. Perhaps it is time to preach the mantra of individual rather than collective responsibility. It is only when individuals come to the realization that they and they alone have the final say with respect to themselves that we will begin to see true responsibility.

And reality is individual responsibility. The African American women infected with AIDS acquired the disease through their own actions. Unprotected sex and shared needle use are responsible for well over 95 percent of new AIDS cases. Protected sex and individual needle use would go a long way toward reducing the incidents of AIDS. Such measures are an individual's responsibility.

And it's not just AIDS. The Japanese citizens killed in World War II had it within their power to prevent their leaders from doing what they did. For whatever reason, they chose not to, and they suffered the results of that choice.

This cannot be overemphasized. Everyone everywhere is going to have to deal with the results of decisions, made or not made, in every segment of their lives. This is reality. Anything else is delusion.

Knowledge is the key. It is the knowing and understanding and acting. We must understand the motivations of our leaders. We must recognize the genetic predisposition. We must confine and direct the innate longing to lead. We must control our alphas.

One of the most amazing aspects of dictatorship is the ease with which they are set up. There are always those in the wings eager and anxious to facilitate the establishment of a single dominance system. The concept seems to strike some deep need in the human animal. These individuals feel better, feel more at ease when one line of authority is established, one line—one person on top and everybody in their place underneath.

Such organizations of individuals are absolutely essential to dictators. If one is to control a million or a hundred million or a billion people, one cannot do it alone. One needs organization to watch, to report, and, when the dictatorship is challenged, to act.

Imprisonment of dissidents is common; torture and execution are used to maintain control; laws, rules, and regulations are established; authority is asserted; and groups are dominated. To be a part of such a linear control system is easy for many people, especially if they see it as a way to validate their position in the dominance hierarchy. Such systems also establish a sympathetic path to achieving additional power.

Dictators and dominance have had an inordinate amount of influence on humanity. The slaughter of the multitude goes hand in hand with dictators and dominance. While it is true that, at one time, the ability of a dictator to decree death was limited, the last six thousand years has seen a rapidly expanding capability of the dictators to dominate and destroy.

But it has only been during the last three or four hundred years that slaughter on any kind of substantial scale has occurred, and only during the twentieth century that a truly massive butchery of our fellow humans has taken place. The potential was always there, but it was the triumph of technology that allowed our alphas to perform so heroically.

Tens of millions of people died in two world wars. In addition, many millions more have died in a number of utterly failed social experiments. Lives essentially lost for nothing, for no good reason, for no purpose other than that a single individual could demonstrate his dominance.

We can learn from this. There are those who seek power and control over others. They are completely ruthless, they are without remorse, and they are totally committed to alpha status. They will do whatever they need to, whatever they must do to succeed. They cannot, short of killing them, be stopped. They can, however, be controlled, and it is important that followers understand that they have that control. Because of their impact on humanity, our crop of dictators will be dealt with in some detail, but first, a short digression.

Fred Winters died a while back. Fred Winters was eighty-six and had heart trouble, so it was no surprise when he suffered a massive heart attack and died almost instantly.

Fred was born in the little town of Cactus Buttes (it was supposed to be Cactus Creek, but the post office said there already was a Cactus Creek, so the first settlers settled on Buttes) in 1917. Fred was the only child of Emma and Roger Winters. Fred grew up in Cactus Buttes, living with his parents in a little house on the corner of Fourth and Main. He attended the small schoolhouse down the road aways and graduated just after his eighteenth birthday.

Fred had an interest in mechanics and, even in school, had a reputation for being able to fix things. Even before he left high school, Fred went to work for the only garage in town. Evenings and Saturdays would see Fred busy at the one job he held all his working life.

Fred learned to operate the machinery and make the repairs surrounding farmers needed to keep their equipment and tools in good working condition. Fred was a good, conscientious employee. He came to work on time every day, worked hard, and used his talent and ability to fix problems, first for local farmers and then later for motorists as they passed through the sleepy little town. He read up on motors, transmissions, differentials, and the thousand and one parts that go into making any piece of mechanical equipment. Fred even got a bit of a reputation as a troubleshooter and was sometimes able to solve a particularly difficult problem. Most times he could help, and his few failures were not enough to tarnish his reputation.

Fred left Cactus Buttes only once. On his twenty-first birthday, he took the bus to the county seat to take a physical for the draft. In the course of his medical examination, he was informed that a problem with his feet made it improbable that he would ever be a soldier, and he was sent back to Cactus Buttes to resume his career as a mechanic.

In 1962, Fred's father died, and Fred became the family's breadwinner. He continued to live with his mother in the same little clapboard house he was born in until she died in 1968. From then on, Fred lived by himself in the little house on the corner of Fourth and Main.

Fred retired from the garage in 1990 at the age of seventy-three. He would have liked to continue, but his hands were so crippled by arthritis that he could only put a nut on a bolt with difficulty. He spent more than fifty-five years at the same job at the same place, doing the same thing. Up every day, six days a week, at five thirty, he ate breakfast and read the paper. Fred was at work by seven; he got off at four, went home to supper and to bed with the chickens. Saturday was the lone exception. On Saturday, Fred worked until noon. Saturday afternoon was shopping and the treat of a supper at the local café. Always it was the same—meat loaf and mashed potatoes with the vegetable of the day, coffee, and a piece of apple pie. Apple pie was Fred's one indulgence.

On Sunday, Fred went to church. He would don his one suit complete with white shirt and string tie. He walked to the Community of Christ Church. The church made Fred a tad uneasy. He wasn't sure about all this religious business, having first been introduced to God by his mother and continuing attendance after her death out of nothing more than habit. The whole concept was a bit overpowering for a country boy, and when the preacher really got carried away with fire and brimstone, it grated on Fred's complacent nature.

He sat in the back mostly, mumbling the hymns, standing and sitting in response to those in front of him rather than from any scriptural cues. Fred wasn't sure if he liked church and was relieved to escape the service at its conclusion and reap the modest rewards of the pious. He liked the small talk with the congregants he had known and respected for many years. Church was a mixed bag. Fred went because he always had and because change, any change, was difficult if not impossible. He was not sure, however, if he liked it.

After retirement, Fred liked to sit on the front porch of the little clapboard house and watch the traffic on Main Street. He had his radio to listen to and a cup of coffee to sip as he watched the world go by. So he was content. Every once in a while, someone, remembering Fred's talent, would bring a problem around, and Fred would try to solve it. As Fred grew older, the music from the radio got louder because he was losing his hearing.

Then one morning, the next-door neighbor noticed the silence and, after a while, called the constable. Knocks on Fred's door produced no response. And so the constable went inside just to check things out. (Never in his life had Fred ever locked his doors.) And there was Fred lying on his bed as if asleep but actually dead of a massive heart attack.

Fred was like many of us today. He worked hard all his life. He was kind, gentle, unselfish (he left his considerable wealth to a local charity), and ethical. He did small good. The many farmers and motorists he helped over the years knew him as a master craftsman. The citizens of Cactus Buttes saw him as a valuable asset to small-town life and were genuinely sorry at his passing.

But Fred was not an alpha. He was content, modest, and ambitious only within a small clearly defined window. Many times the owners of the garage (grandfather, father, and son) urged Fred to take the position of lead mechanic, of foreman, even offering him the chance to manage the garage.

Fred would have none of it. He liked machines and tinkering. He had no desire to boss others or to "run things." And so Fred Winters spent fifty-five years doing what he was comfortable with, what he liked, and, with his passing, leaving only a small mark upon his world. Fred Winters was not an alpha.

CHAPTER 12

Joseph Stalin was an alpha. Joseph Stalin, by some very good estimates, murdered (or rather was responsible for the murdering of) over sixty-one million people. Joseph Stalin had a great and terrible impact on mankind. He and his minions carried out the single greatest inhuman act ever committed by humans upon humans. Actually, Stalin did not do all the work himself, although there is evidence that he took an active part in a number of murders.

To put this horrific slaughter into focus, a little math is needed. In his approximately thirty-six-year reign of terror, Stalin was responsible for the deaths of 61,000,000 people. That is 1,694,000 people a year or 32,585 a week or 4,655 a day. This works out to almost 3 1/4 people every minute, a little over 18 seconds per killing. Every day, day and night, holidays and weekends, without exception, somebody died. It is obvious Stalin could not have done the job by himself; he had to have had help.

Stalin and his henchmen killed 61,000,000 people because they considered these individuals to be a threat to their dominance structure. Oh, there was the occasional vendetta, the random hatchet job, the "I'll show you" denouncement, and in many cases, the regime killed not only the perceived threat but also the threat's family. But mainly it was a perceived threat to dominance that got one axed.

The dominance hierarchy did not discriminate as far as method goes either. Many were shot, but shooting takes time, and then there were all the bodies that needed to be disposed of. More were worked to death in the gulags, but most were simply starved to death.

The kulaks were wealthy or not-so-wealthy landowners. They were farmers, and they fed Mother Russia. They were conservative, clannish, and powerful. Because they controlled the grain that fed the country, Stalin considered them to be one of the greatest threats to his dominance. So he took away their grain and their livestock and their fruits and vegetables and everything else the kulaks grew.

The idea was that the state would take the ill-gotten assets of the avaricious kulaks and redistribute them to the poor but deserving masses. In a deliberate policy intended to break the back of this powerful group, Stalin decreed that the kulaks would be allowed to keep nothing, and so the kulaks starved.

Everything edible was taken away from millions of people, and as a result, millions of people died. Periodically, the goon squads would go through the ever more deserted towns and hamlets of Russia and check for progress. If the peasant population looked sufficiently emaciated and the body count was up, then the government was happy. If some farmer and his family looked suspiciously healthy, they were shot, the rationale being that they obviously had something to eat, and since all food was to be turned over to the single dominance system, they were committing a crime against the state. The punishment—execution for everyone, even the babies.

The sixty-one-million figure is not just a guess. This figure has been authenticated by Professor Rudolph J. Rummel, a political scientist at the University of Hawaii. The results of Professor Rummel's research are available in the newsletter of the Center for the Study of Social Conflicts at the University of Leyden in the Netherlands. What the good professor is talking about is the killing of unarmed, nonresisting individuals during times of peace.

Lenin said there was a need for "revolutionary violence against the faltering and unrestrained elements of the toiling masses themselves." *Faltering* and *unrestrained* are euphemisms for those who refuse to go along with the communist program. There were millions of faltering, unrestrained individuals in Russia.

Communism has not worked and will not work. We are social animals. All social animals have dominance systems as part of their behavior pattern. This is a genetic predisposition and cannot be overridden. It can only be influenced by cultural mores. There is a very good reason why each communist country begins its reign with an orgy of butchery. The alphas, the dominants, are not interested in egalitarianism. They are not interested in equality, justice, or the betterment of mankind. They want power—total, complete, and unabridged power. The easiest way to accomplish that goal is to murder anyone who displays the slightest potential to challenge that power.

Stalin came to power in Russia and murdered millions. Mao came to power in China and murdered millions. Pol Pot murdered millions. Tito murdered hundreds of thousands. Castro murdered tens of thousands (he would have killed more, but Cuba's total population is less than ten million). It is simple; it is straightforward. We are a dominance-oriented species. These are dominance-driven individuals, and what we see in their behavior is the unrestrained execution of that dominance.

Russia was not alone in being a communist country. It was not alone in being a dictatorship or in repressing its own people or butchering them. China, too, presents us with a fascinating display of dominance, delusion, and man's inhumanity to man. China—huge, ancient, decrepit, and falling, until recently, further and further behind in the technology race. This country, at one time, decided to allow the people to criticize the communist government, and in 1957, the Hundred Flowers Campaign was born. The idea was to let one hundred different flowers bloom to let a hundred or even a thousand or more ideas bloom. In short, the communist party would listen to the people.

One doctor suggested that the uneducated peasant soldiers, who had been turned into party secretaries and made hospital administrators by the government, be replaced by competent hospital administrators. This was a reasonable suggestion inasmuch as these peasants had absolutely no idea of how to run a hospital. They routinely made decisions that cost money, time, and lives while doing absolutely no good at all.

Then command and control kicked in. The authorities were at first amazed and then shocked that anyone had any remotely reasonable suggestions to make, and the Hundred Flowers quickly faded. The good doctor was persecuted, prevented from practicing medicine, and as if that wasn't bad enough, his wife, also a doctor, was banned from practicing medicine. His son, one of the smartest students in all of China, was sent to the countryside to perform manual labor.

It is typical that the powers that be gave no thought to the fact that their decisions deprived the Chinese people of two very good doctors and the potential for another one. These individuals, who claim to have the interests of the people at heart, have no qualm or quibble about depriving the population of someone educated and trained to take care of that same population. Nothing stands in the way of these alphas and their quest for power.

The story of the doctor is not unusual or uncommon. And many a Chinese citizen found, to their dismay, that the leadership would brook no criticism of the party. But the Hundred Flowers Campaign was small potatoes compared to what took place a year later in 1958. Mao Zedong had an idea. He would confiscate all the remaining private property in China and reorganize the entire country into communes. He was absolutely convinced that the great energy of the masses would transform him from an absolute dictator of a large, backward, poor, and rural country to a dictatorship over a modern, booming, technologically advanced country. He failed, and in the process, over sixteen million people died.

No better illustration can be made for the idiocy of this program than Mao's plan for grain production. He became obsessed with grain production. This, he maintained, was the only way to measure the success of the farm plan as handed down by the command structure. He decreed that all rural communes would, from this time on, produce grains because grains were the key link in food production. This meant not growing all kinds of fruits and vegetables as well as essential oils from soybeans and rapeseeds. Production of these items was reduced by as much as 50 percent in order to meet the top-down demand for grain. The resulting deficiencies in diet meant that millions more Chinese would suffer from the debilitating effects of malnutrition.

For an excellent book on the problems of a socialist command economy, I would recommend *China: Alive in the Bitter Sea* by Fox Butterfield. This volume gives the reader some idea of the more practical problems of a command economy. It is also a reader, for those who see, on the nature of human nature and the perquisites of power.

Mao Zedong was an alpha. Mao Zedong, by Professor Rummel's estimates, murdered over thirty-eight million people. Mao Zedong had a great and terrible impact on the people of China. He and his henchmen were responsible for the second greatest inhumane act ever committed upon humans by humans.

It is no coincidence that Stalin and Mao committed these monstrous crimes in the name of a perceived ideal state for humanity. And that state was communism. But then we often do the worst of deeds for the best of reasons.

Adolf Hitler was definitely an alpha. Hitler and his underlings were responsible for the murdering of over 20 million people. The Japanese, under Tojo, killed almost 6 million people. Pol Pot killed 2 million people. The list goes on and on. Best estimates for the century just past: over 143 million people were murdered by their own governments, mostly during times of peace.

The killings do not include those lost in wars. In fact, the victims of all the wars of the twentieth century were a little over thirty-five million. Governments have deliberately killed three times as many individuals as have died in all the civil and international wars fought during the last hundred years.

And the killing continues to this day. In Laos, thousands die every year as the communist government tightens its grip on that poor defenseless nation. North Korea and starvation go hand in hand. Hundreds of thousands have died, and hundreds of thousands more will die in the next few years because the dominance structure in that country views them as a threat. But many more will die because the government doesn't care if they live or die as long as the elite retains control.

Fidel Castro was an alpha. Fidel Castro was responsible for the murder of tens of thousands of Cubans. Anyone and everyone who posed the slightest chance of challenging his dominance was either killed or sent to prison. Castro was not interested in the welfare of Cubans, nor was he interested in the improvement of Cuban society. Castro's sole interest was dominance. In this respect, he differs not one whit from his predecessor, Batista.

Castro claimed he had the best interest of the Cubans at heart, but reality intrudes. He claimed that Cubans have given up all their freedoms for "education, health, and science." It is true that he has educated the Cubans, but he has educated them to the poverty of the socialist system. True, he has educated more doctors than any other small island country in the Caribbean, but his socialist system is so poor that the country cannot afford the medicines necessary to treat the sick. His socialist system is so poor that the average Cuban shows signs of diseases consistent with a diet poor in nutrients. Lest anyone believes that it is the blockade by the United States that causes this problem, that is not the case. There are lots of countries ready and willing to sell to Cuba any and all items needed to give the average Cuban adequate nutrition. Fidel Castro could not afford to buy these items from anybody, let alone the United States. Anyway, as long as he was able to preserve his dominance of the Cuban people, he did not really care.

There is yet another side to the Cuban coin. Fidel and his cronies did not share the privations and frustrations of the average Cuban. No, Castro lived high on the hog with lots of meat and chicken in his diet, lots of veggies, too, if he wanted them, and ice cream for dessert. This class delineation in a supposedly classless society is common among dictatorships, so Fidel Castro's behavior was not that unusual.

Castro was not alone in his paranoia of power. Pol Pot, too, solidified his dominance through the murder of two million Cambodians. His goal was a bit different from that of Castro's in that his vision of utopia was everybody down on the farm. All would till the land, and Cambodia would become the breadbasket of Asia, and Pol Pot would be head baker.

Lots of Cambodians could and did farm, but many more were shopkeepers, tradesmen, professionals, teachers, or worked in other educated occupations. These individuals Pol Pot slaughtered on a massive scale. Doctors, engineers, civil servants, and educators were all murdered—a whole generation of educated and productive individuals were made victims of a madman's lust for power. Pol Pot did not murder all the professionals. He kept enough doctors around to treat the elite. He kept enough teachers around to teach the elite, and enough civil servants to turn the few wheels of government that were left.

Everyone else was sent to the country, given nothing, forced to work sixteen-to-eighteen-hour days, flogged, beaten, starved, and, in the end, murdered. Piles of skulls, pyramids of skulls, huge piles of bones were all that was left of two million people killed to satisfy one man's need for power and dominance.

Pol Pot was an alpha. It is critical that we identify and understand alphas. We cannot prevent their existence, but we can control them. The Cambodian people did not understand that and have paid for their ignorance with their lives, their wealth, and their freedom. Cambodians welcomed Pol Pot as a viable alternative to the existing monarchy. Only a very few recognized the possibility that Pol Pot would be infinitely worse than their present master, and they were the first to die. Anyone who spoke out against Pol Pot was murdered. Anyone who posed the slightest possible threat to Pol Pot was murdered. The victims—mostly men—were shot, hanged, poisoned, and beaten to death. The Cambodians did not understand the motivation of Pol Pot; they paid the ultimate price for that lack of understanding.

As I have said, what is most discouraging about the establishment of dictatorships is the relative ease with which they can be instituted and maintained. Castro had no trouble rounding up enough willing participants in his scheme to dominate the Cuban people. The core of his power came from those individuals who helped him overthrow Batista. But Castro was able to organize and implant a rigorous and pervasive system of spies and informants that operated right down to the block and building level. Those normally on the bottom rungs of social life jumped at the chance to lord it over the populace and share even a little bit of Castro's power.

Stalin, too, as we have seen, had no problem establishing one supreme dominance system throughout Russia. A huge number of willing participants were able, not only to carry out his plan of murder, but also to meticulously document the acts—individuals willing to follow up murder with the death or imprisonment of the victims' family. They were also more than willing to engage in the brutal interrogation of prisoners in order to unravel the perceived threads of Stalin's self-induced psychosis.

It is all too easy for a dictator to set up a dictatorship. There are far too many people willing, even eager, to assist in the setup of a single dominance system. There are far too many people in a given population more than willing to intimidate, torture, imprison, and kill their fellow citizens for no other reason than the fact that committing such crimes gives them a sense of power over their victims.

The question must be asked. Are these individuals simply misguided fools paving the road to hell with their perceived good intentions, or is it possible that these despots know exactly what they are doing? Is it possible that their goals have nothing to do with liberty, fraternity, and equality?

Russia wanted desperately to become a great nation. It tried a great experiment. Sixty-one million people sacrificed their lives to the great experiment, but it did no good. The great experiment failed, and the question must still be why. Why was this huge country so rich in people and resources unable to become a great nation? Was it because Russia was unable to understand and tame its alphas?

Dominance is one of the big disadvantages to civilization. Dominance on a massive scale can only occur in civilized societies. As we have seen, even today, tribes in the Amazon jungle or the New Guinea highlands are limited in their ability to exert influence over neighboring tribes. They have neither the technology nor the capability to overpower and dominate neighboring rivals.

This is the way it has been for eons—each tribe, each group, with roughly the same capabilities and potential. Civilization changed all that. Civilization gave us the ability to organize and assemble large groups of warriors for the purpose of domination. War and dominance have always been with us, but it is only the civilized countries that can prosecute a war on a massive scale. This is something we seem to have a great liking to do. This is something that says little good for civilization.

Alphas are not all. They are a large part of but not the total answer, and controlling them is not the total solution. As we have seen, the human animal is singularly plagued by delusion. This makes the dictator's path easier, and it sheds light on why we behave the way we do.

We like to think that we are nice, kind, gentle, and honorable. We view transgressions such as rape, assault, and murder as anomalies. And yet we have no trouble fielding an army ready to kill others, to assault our fellow man in the interest of our culture. We have no trouble legally taking the life of individuals who have taken the lives of others. We can and will do things to others that are not nice but are within the parameters of our culture. It is not a case of us being nice, kind, gentle, and honorable. It is rather a situation that the culture establishes the parameters of acceptable human behavior, and sometimes that acceptable human behavior is really, really bad.

Jedwabne, July 10, 1941. A small town in Poland is about to experience a drastic population reduction. One half of the citizens of Jedwabne will murder the other half. The story does not bode well for the nature of human nature.

At the start of World War II, Russia and Germany divided up Poland. Jedwabne ended up on the Russian side of the division. In June 1941, Germany attacked Russia and "liberated" Jedwabne. One of the first questions asked by the newly liberated gentiles of Jedwabne was "Is it permitted to kill the Jews?" The Germans said, "Yes," and so one half of the town killed the other half.

Fifteen hundred Jews died at the hands of their neighbors and fellow townsmen. They were burned, drowned, knifed, but mainly beaten to death. This was a particularly satisfying and personal way of disposing of someone. Babies were pitchforked, small children were thrown alive into bonfires, and men and women were burned alive in the flames of their businesses and homes. When the killing subsided, less than a dozen Jews survived in Jedwabne, and the town's population was one half of its pre-pogrom size.

The killers of the Jews were ordinary Poles—Catholics, civilized, educated, and mostly law-abiding individuals. They were offered an opportunity to do something hideous without fear of retribution, and they took it. The thin veneer of civilization slipped, and the true nature of man revealed itself. Jews were strangers. They ate funny, looked funny, and, most of all, behaved funny. It was easy for the men of Jedwabne to kill them—simple, no risk, really, for the Jews had nothing with which to defend themselves. It was easy for the women to watch and encourage and beam with pride as little children were thrown into the fire. The Poles of Jedwabne acted normally, naturally, and with a degree of finality—the same thing any good human would do when confronted by the hated enemy.

A small diversion. It has to do with hatred and killing and Jews and dominance. So I suppose it has something to do with our search for reason in the murk of man. It may shed some small light on the Poles of Jedwabne and the terrible deaths of 1,500 Jewish men, women, and children.

For the last two thousand years, Jews have seen incident after incident like that of Jedwabne. Some are smaller; many are larger. They demonstrate persecution on a global scale. No one likes the Jews, and no one knows why. The answer from a religious standpoint is quite simple. Jews are persecuted because of their religious philosophy.

When we view various religions dispassionately and objectively, it becomes apparent that they are, for the most part, dominance systems. The pope is infallible. He is the one dominant male in a chain stretching from St. Peter's to every church, parish, convent, and monastery in every country in the Catholic world. And just like any other dominance system, the pope has the final say over everything. Mormons, too, have a strict dominance hierarchy. Muslims have a religion built upon dominance with the grand ayatollah having an inordinate amount of power. And so it goes; each religion is built upon a dominance system with the structure of the system assured by the fact that the religious rulers hold our souls in their hands. And so we must do as they desire or go to hell.

If you were to ask any honest Catholic, they will tell you that Catholics and only Catholics are going to heaven. Everybody else can go to hell. If you press a devout Mormon as to whom they will share eternity with, they will tell you Mormons only. If you ask a Muslim who is bound for heaven and who for hell, he will tell you that only Muslims enter heaven. This is the stranglehold the clergy has on their adherents. Do as we say, or you will be punished for eternity—not a very desirable future for delusional individuals.

Faith is the key. "Keep the faith, baby" is the mantra of religion. Everything depends on faith and believing. The key to clergy survival is to blackmail believers into accepting religious dominance, else they lose their eternal souls. Faith is the key. It is faith, not deeds, that drives the religious authority to do what is necessary, to do whatever is required, whatever they have to, to survive. And so the religious rulers have, down through the years, committed some horrendous deeds in order to retain the faith of the populace, all built upon the dictum that faith is more important than deeds.

Not so Judaism. Jewish philosophy hews closely to Micah 6:8: "And what doth the LORD require of thee but to do justly and to love mercy and to walk humbly with thy God." Jewish religious philosophy emphasizes deeds rather than faith. Notice the quotation does not say "my God" or "our God" but "thy God." Jews do not proselytize for the very simple reason that Jews do not claim to have a lock on religion. Jewish religious belief maintains that all who do good work on Earth will share eternity. This is why Judaism has been a persecuted religion for the last two thousand years.

Jewish religious views are an absolute anathema to individuals bent on domination through the blackmail of faith. If Jewish religious philosophy ever goes mainstream, the religious dominance hierarchies will come tumbling down, and the purveyors of purgatory know this. This is why Jews must be purged. This is why pogroms occur. This is why laws and rules and regulations are instituted in an attempt to limit and contain the Jewish blasphemy. Dominant individuals will do whatever is necessary, whatever they have to, in order to maintain their status.

Which is more important, faith or deeds? Most religionists will tell you that both are equally important, but reality intrudes. I once asked this question of a pair of Mormon missionaries and got the standard answer. The following comparison puts the lie to their reply.

Suppose you have this man, as deep and black a scoundrel as ever there was. He robs from the widow and orphan; he cheats his partner and his wife and his mother. He is, in short, the most despicable of men; never in his life has he done a good deed. On his deathbed, the man genuinely regrets his past life, feels true remorse, and begs forgiveness. He embraces Jesus and the Mormon church and is truly repentant. What will happen to this man?

"Well," the pair said, "he will go to heaven."

"Suppose," I said, "you have a man who has done nothing but good deeds all his life but because he lives on a remote island in a remote ocean, no one has come by to explain Jesus and Mormon and going to heaven?"

"Well," they said, "when he dies, he will be reincarnated so that he can be introduced to Jesus."

So we have one man who did nothing but good deeds, and he cannot get into heaven on one hand, and on the other, a man who has only done one good deed and that to his own benefit, and he is going to heaven. What's more important, faith or deeds?

The church will tell you that both are important, but this is not the case at all. You see, it is by faith and faith alone that religion lives and dominates. And this is the goal of religion: to establish a dominance hierarchy to the benefit of religious officials.

This is not just a philosophical discussion. When faith becomes more important than deeds, it can lead to all kinds of inhumane acts. Suicide bombers place faith far above deeds. The killing of women and children is desirable as a way of validating one's faith. But Islam is not alone. Burning at the stake is about as despicable a deed as one could imagine, but the Catholic Church called it an act of faith. All religion is ultimately based on faith. No religionist can, by a series of simple scientific statements, demonstrate the validity of their religion. Religionists cannot even demonstrate the existence of God. Faith is all they have.

Deeds are more important than faith. I don't care if you believe that the moon is made of green cheese and the craters we see are really bites taken out of the lunar landscape by space rats. What you believe has no bearing or impact on anyone. What you do can and does have a great deal of impact on those around you. Deeds are far more important than faith, and the Jews realized this a long time ago. It is this religious philosophy that causes the Jews so much pain and anguish. And most Jews do not even realize that the concept of deeds being more important than faith is the cause of many of their problems.

This is why the Jews of Jedwabne died. The Catholic Church sees Judaism as a real and profound threat to its dominance. It must be contained, diminished, and, if possible, eradicated. For centuries, France has persecuted the Jews, killing thousands in 1321 and finally expelling all Jews from France in 1394. Spain, too, expelled its Jews and, in the process, lost most of its doctors, scientists, and teachers. The loss of all that intelligence still impacts Spain to this day. Then, of course, there was the Holocaust in Germany and in countries occupied by Germany during World War II. Six million Jews and absolutely eons of talent, intelligence, and knowledge are gone. A terrible loss and all because Jewish religious philosophy prefers deeds to faith. It is a terrible indictment of humanity. So many have died for dominance, and the dead do not know why.

CHAPTER 13

We are evolved animals. We lived and changed and evolved on the plains and in the seas of East Africa some seven to eight million years ago. We were, at best, a marginal species, the unlikely result of a series of stopgap, makeshift changes in response to fluctuations in climate and habitat. We were, until a million or so years ago, rare in the pantheon of African life. Then we conquered fire and, with its help, spread around the globe. We were, at that time, predators, and our densities were those of a predator. It was only later when our populations inhabited the known world that our brains really started to expand, and this occurred in response to competition with ourselves. We are here today as a result of chance and circumstance. We could very easily have gone extinct or, had things been a bit different, still be crashing about in the woodlands of Africa.

We are evolved animals, and we behave in ways that demonstrate that evolution. We love, hate, fight, and mate all in response to a greater or lesser genetic predisposition. It is only through evolution that we can begin to understand why we are what we are and why we do what we do. This is why the concept of creationism has been dealt with in such detail. Creationists would have us made from whole cloth with no past and no way to come to grips with human behavior.

The elite, too, present a false picture of humanity—all sweetness and light with no really good explanation for man's truly inhuman behavior. All is nurture, claim these self-deluded zealots. If we treat everybody "nicely," then everybody will be "nice." It works just often enough to keep the elite hoping and praying that they are right, but not often enough to lead to any realistic results.

We are dominance-oriented animals, and this predisposition shows itself in everything we do. Cultures are dominance systems built and maintained to provide a logical, if not rational, program for delineating relative position within the community. We are a mixture of alphas, betas, gammas, all the way down to the lowest omega man. Most of us range in the middle of the pack—thetas, kappas, down to pis, and sigmas. We do our thing, make our contribution, live a more or less uneventful life, and pass on, having done small good and leaving happy memories.

We see dominance everywhere. We see dominance in dictators—Stalin, Mao Zedong, Hitler, Pol Pot, Caesar, and as far back in time as one would care to go. Death was on an unbelievably massive scale, robbing the world of the potential of millions and millions of lives. How many Mozarts? How many Einsteins? How many individuals with the potential to do great and good things for mankind were lost in the Holocaust? In the purges? Or in the Great Leap Forward? It is a sad reflection on humanity when one realizes that we are our own greatest enemy, that only man has it within his power to destroy man. And it is all because of dominance—dominance and delusion. Uncounted thousands of supposedly sane, rational men and women extolled the virtues and values of the great Russian social experiment even while millions of people died a slow, painful death from starvation.

But are these people really as delusional as I have made them out to be? Unfortunately, the answer must be yes, even for those in the know.

Consider Walter Duranty. Duranty was the *New York Times* reporter in Russia during the early thirties. It was during the early thirties that the first rumors of starvation started to leak out of Russia. As we have seen, Stalin had deliberately developed a policy of starving the farmers to death in order to remove them as a challenge to his authority. Duranty was there. He saw the carnage.

He not only refused to report it, but also he denied its existence. Time and time again, he wrote in the *New York Times* that there was not starvation in Russia even though he knew that more than twenty-five thousand people were dying every day. He even won a Pulitzer Prize for his reporting.

It is hard to understand this behavior. Duranty was not deluded in the sense that he did not know what was going on. He did. He knew and lied for a bigger reason. Duranty was convinced that communism was the wave of the future. He just knew that deep down we were all nice even though he was not. His rationale for blatantly lying to the American public was simply that he just knew that communism was right and proper and the salvation of mankind, that it was perfectly proper to lie about it. Anything to bring the revolution about. All was acceptable and permissible because the end justified the means.

Dominance drives us (many of us); delusion blinds us (most of us). These are the two defining characteristics of the human animal. We share a dominance predisposition with all social animals. But delusion is a singularly human condition. We are special, we are unique, we are different, we are better, we are kind, we are good, and we are a truly grand culmination to the evolutionary process. We may not be gods, but we look like gods. They fight over us and even make sacrifices to us for our own well-being. We are so important, so critical, so absolutely essential to the universe that at one time the whole of the world must be expunged except for a handful of humans and a few pets. These are our delusions.

Delusions cause problems, real problems, huge problems. Religions are supposed to be bastions of moral behavior. Human belief in religion has been pointed out time and again as a truly specific human characteristic. Moral behavior, many aver, is a defining human quality, and for a large percentage of the population, this is probably true.

The problem lies not with followers but with leaders. It is with these individuals that we see the seeds of dominance and the disappearance of morality. One need look no further than the Catholic Church. Once it gained ascendancy in the 600s, it began an ever-increasing drive for dominance. This culminated in the Inquisition where hundreds of thousands of individuals were tortured and burned at the stake because they posed some perceived threat to the dominance of the church. I cannot imagine a more horrible death than to be consumed by flames, to be unable to get away from the searing heat as the flesh is roasted from the living bones. The church called this an act of faith.

We see the same problem with Islam today: total disregard for human life, absolute conviction of infallibility, and a willingness to use ignorant, gullible, and superstitious individuals to achieve the desired goal of absolute world domination.

It is the alphas that cause the problems. It is the dominants that direct and instruct and indoctrinate the followers to do bad things for a so-called good cause. It is the dictators that kill and maim and torture and destroy. It is the kings who have divine rights among which are the rights to take our lives at their whim, especially if they think we pose the slightest threat to their dominance. They also kill people just to demonstrate that they can and they are not to be messed with. It is with the alphas, always the alphas, that we must contend. We must know them, understand them, and, most of all, deal with them.

We cannot do without alphas. They are necessary and important to our social fabric. Without structure, there would be only chaos, and even then, alphas would dominate. What we must learn to do is to control our alphas. It can be done. Alphas will operate within the confines of a culture, but only if they are forced to do so. They would prefer to exercise their genetic predispositions free of restraint, but they can be made to conform. We have already seen examples of this in the big-man ceremony and in the potlatch ceremony. In both of these instances, alphas do good for the community as a whole in order to achieve the perceived status that they aspire to. Alphas will follow cultural dictates. They will respond to the demands of the followers. This is accomplished through democracy.

The dictionary describes *democracy* as (among other definitions) a rule by majority, or rule by the people either directly or by a representative. But what all this really means is that democracies control their alphas. If an individual wants status and dominance within a democracy, he or she must act in ways that the majority approves of. This, of course, means that dominants cannot murder potential rivals to improve their chances of success. Alphas are limited by both law and custom as to how they can achieve their goals. In addition, after they have achieved dominance, they must continue to behave in ways that the followers approve of. This happens only in democracies.

If one were to line up all the countries on Earth and rank them both by the degree of democracy the countries afford and the value of personal wealth within that country, there would be a very close one-to-one correlation. Democratic countries are rich countries.

There are a few exceptions. In the Middle East, a few Arab states export oil and use the proceeds to finance a luxurious lifestyle. This is a temporary condition, however. Eventually the oil will run out, and economics will take over. Singapore is a dictatorship, albeit a "benevolent" one, and Singapore is very rich. But the dictatorship is political while the market system is capitalist. So even some dictatorships can and do create wealth. But by and large, the more autocratic a country is, the poorer it is.

Africa is a continent of poor countries, and with one or two exceptions, Africa is a continent of dictatorships. South Africa is a democracy. True, at one time, it was a limited democracy, actually disenfranchising a majority of its citizens. This has, however, changed. South Africa is a rich country; it remains to be seen if it continues to be so. Almost every other country in Africa is poor, despotic, and criminal. Even Kenya, a British colony that was left at the end of colonial rule with a sound, solid economic and political system in place, has managed to find its way into the realm of despots and dictators.

Idi Amin died on August 16, 2003, in Saudi Arabia, in bed, in comfort, with his boots off. Amin ran Uganda for a number of years, and his reign is one of several that stand out as an extreme example of dictatorial, despotic, and cruel rule. He ate people's hearts, literally, claiming to acquire the capabilities of his dinner guests. An alpha, a dictator, a totally unscrupulous individual who stopped at nothing to gain and maintain his position as number one Ugandan. Amin was a dominant.

The people of Uganda paid for the actions of Idi Amin with their health, with their wealth, and with their very lives. Followers will inevitably be held responsible for the actions of their leaders. Followers may not want to be held responsible, may even feel they should not be held responsible, but responsibility is the very essence of leader-follower relationships. Any individual or group of individuals acknowledged as leaders, willingly or not, will impact directly everybody who follows them. It is, therefore, essential that leaders be chosen rather than imposed. Followers must understand that the final responsibility for their leader's behavior rests with them.

Liberia is a somewhat unique but enlightening example. In the 1820s, American abolitionists put together a plan to send freed slaves back to Africa to establish their own country. Liberia was carved out of the jungle of Africa without much consideration for the resident natives in the pursuit of this goal.

Now one would think that having tasted slavery, the freed slaves sent back to Africa would appreciate the negatives of subjugation, but not so. The newly freed slaves promptly enslaved the natives and set themselves up as lords of the manor. The colonists took over the government and in 1847 established a so-called Republic of Liberia. It's been all downhill ever since.

It got so bad that in the 1930s the president of Liberia, a country colonized by former slaves, was forced from office because he was selling uneducated countrymen into slavery to work on cocoa plantations. The slaves who were looked down upon and despised in the United States, in turn, took on the robes of despisers in Liberia. The problems continue today.

Not to say that the country hasn't tried. There have been a number of what most describe as fair and honest elections. The problem arises once the president is installed. Dishonesty reigns, greed begets, and evil triumphs. Power corrupts, and newly elected presidents drink deep from the well of immorality.

Liberians do not have any control over their alphas. It is the responsibility of the Liberians to control their dominants. When they do not, they pay for the behavior of their alphas with their wealth, their future, and, in many cases, their lives. It is not enough to elect leaders; followers must also force their elected officials to behave in ways that benefit the people. Such behavior is not natural or normal, the inclination being to do what is best for self at the expense of others. The goal is the accumulation of wealth, the exercise of power for power's sake, and to hold on to power by any and all means. The Liberians must learn to control their dominants. The alternative is murder and mayhem and poverty on a massive scale. This dictum applies to every country in the world.

Just how badly dominance-oriented individuals can muck up a country is exemplified by Robert Mugabe and his dictatorship over Zimbabwe. In 1980, Mugabe took over Zimbabwe, and the results of his twenty-three-year term is corruption, starvation, and murder on a huge scale. A new tool in Mugabe's arsenal is rape. Girls as young as ten and twelve are gang-raped in front of their parents and relatives while the witnesses are required to chant praises to Mugabe. This is the established punishment for voting in rigged elections for the opposition.

Mugabe also uses starvation to keep his subjects in line and under control. At one time, Zimbabwe was a regional breadbasket. The country even exported grain. Today, millions of people in the country face starvation as a result of two Mugabe policies. In the first of these idiotic policies, Mugabe has forced the White farmers who produced the gains from their lands and has given the farms to the Blacks. He has done this without compensation. Most of the farms now lie fallow because the Blacks either do not want to or cannot farm them. As a result, Zimbabwe has to import grain from international aid organizations.

In a second policy of idiocy, this imported food is seized by the government and given only to Mugabe loyalists. All others are forced to buy grain on the black market where a day's supply of food costs more than the average Zimbabwean makes in the same day.

The result is starvation, and current estimates show that two to three million people will die in the next few years from this cause.

Africa, poor Africa. First, there were the monarchs—kings and queens, bad and worse, selfish and contemptible. Then came the dictators—mean, tyrannical, and despicable. Africa has experienced six thousand years and more of the most horrific, murderous, uncivilized, and barbaric behavior ever inflicted on any continent. A situation that for the most part continues to this day.

Africa demonstrates the reality of the human animal. Yes, there are those who say it was colonization that did Africa in. But hatred and animosity far outdate European incursion. The Tutsi have hated the Hutu since time immemorial. The only thing European intrusion did was to provide the combatants with the technology with which the natives can slaughter each other on a grand scale. Bows and arrows limit murder; flint knife butchery is slow and dangerous. Guns, even in the hands of children, kill massively, and machetes lop off hands and feet with amazing dexterity. The hate was always there. Civilization, in its infinite wisdom, allowed death to be displayed in hundreds of thousands rather than in hundreds.

It is not as if Africans did not know better. In Northeast Africa six thousand years ago, there arose a civilization the likes of which the world had never seen. Pyramids and temples and statues were built on an awesome scale. The unification of Upper and Lower Egypt under Narmer presented the world with the first stirrings of wealth, surplus population, and the trappings of power. One man, not just alpha over a clan or even a group of clans, but dominant over a vast population. Division of labor on a sweeping scale. They were no longer simple hunter-gatherers but skilled craftsmen in many occupations. Bakers, builders, businessmen, traders, sailors, masons, and engineers—all required for the work at hand. Great civilizations do great things, and Egypt was a great civilization.

And Egypt wasn't just a flash in the pan either. Egyptian civilization in one form or another stretches in an almost unbroken line down through the years to today.

That being the case, why do we not see any emulation of Egypt by any other social groups in Africa? Nowhere else on the continent is there the likes of Egypt. No huge pyramids, no temples, no statues, and, most importantly, no civilization capable of such activity. The rest of Africa remains mired in savage, primitive, barbaric backwardness. The question must be why. Why is Africa today a place of murder on a massive scale, of fraud, of greed, and a continent justifiably called Dark?

Africa is the way it is today because it has never been able to control its alphas. Dominant individuals have usurped the leadership, first of tribes, then of countries. Africa displays natural human behavior writ large. Dictators and despots, in response to genetic predisposition, demand dominance. This is common, ordinary, and ongoing. It will not stop until the people of Africa learn to tame their alphas. It can be done. These individuals are genetically driven to lead, but how they lead can be influenced. They would prefer to dominate without restriction and, like Idi Amin, rule absolutely.

But alphas can be made to govern within the rule of law. They don't want to, they would prefer not to, but they will, provided the people force them to. When African followers insist that their alphas rule in the best interests of the followers according to rule of law, this will happen. When the only way alphas can become leaders is by honest and fair elections, then followers will control their leaders. The only way for Africa to remove itself from six millennia of dictatorship and despotism is democracy.

Africa represents the natural state of man—contaminated by civilization. Africans are behaving normally, naturally, using the inventions of civilization to vent their xenophobic predisposition. Left to their own devices, ignored by Europe and America, Africa today would be mired in intertribal warfare exactly the same as they have been for thousands of years. There is a lesson here. Africans behave as our ancestors have behaved for hundreds of thousands of years. We are the ones who are leading abnormal lives, behaving in abnormal ways, and doing abnormal things. We will not be able to improve the situation in Africa by urging the people there to act naturally.

Africa is not alone as a place that absolutely failed to control its alphas. Consider China. China is the largest country, in terms of population, on Earth. More people than you can shake a stick at, as Ma used to say. And unlike Africa, China boasts a history of civilization piled on top of civilization. The history of China consists of dynasty after dynasty—Shang, Chou, Chin, Han—with each supplying a sense of continuity along with improvements to society. The Chinese invented both the wheel and the compass, two outstanding and primary tools necessary to the advancement of civilization.

And yet China was never able to tame her alphas. Even today this is so, and the question must be why. Why is China, a country with a long and illustrious history, still—for all intents and purposes—a third world country? China is just now beginning to feel the positive aspects of civilization. This country has had dictatorship after dictatorship, and no realistic end is in sight. China is an excellent example of the dominance component incorporated into the theory of dominance and delusion. Delusion, too, rears its very ugly head.

China has the largest number of people in the world under one government. One-sixth of all the people in the world exist under a dictatorship, and a singularly peculiar dictatorship at that. It is nominally communist. Now communism says that each should contribute according to his ability and take according to his needs. A fine altruistic declaration, and it would probably work, too, if only we weren't so damned selfish. Communism is supposedly to work by the process whereby the commune is guided by the dictatorship of the proletariat. But every communist leader in the history of the movement has conveniently ignored the "of the proletariat" and concentrated on the dictatorship instead. As a result, like all communist countries, China is ruled by a small cadre of power-hungry individuals who scheme and plot and plan in order to get as high up as possible in the one dominance hierarchy in existence in the country. Communism has become just another method whereby a genetic predisposition toward dominance is exercised by certain Chinese males. And it is the males; all communist dictatorships were or are run by males.

What is interesting about this is that China today is politically communist but economically capitalist. The latest catchphrase in China is "To get rich is glorious." This is hardly a fitting response to Marxist philosophy, but it seems to work. Many Chinese are getting rich.

This state of affairs does tell us one very important thing about the masters of China. They have no interest in communism. Communism, as we shall see, doesn't work, and the Chinese leadership knows that. Capitalism does work, and so China is galloping madly down the road toward capitalism. Why then do these so-called communist leaders continue to cling to the delusion of communism? The answer is power—pure and simple. It is dominance. The knowing is what stimulates them, the knowledge that they can point a finger and literally heads will roll. What drives these individuals is the knowledge that at any time, at any place, they can and will be perceived as the most dominant individuals around. This is why they do the things they do.

When communists take over, heads roll. When dictators take over, heads roll. When one king replaces another, heads roll. And it is always the same. If the individual is perceived as a threat, that individual is dead. Once in power, these individuals consolidate power by eliminating anyone and everyone who presents the slightest possibility of challenging that power.

It is the thought of power, of dominance, that drives these individuals. It is not the betterment of mankind, nor is it the elevation of man that motivates dictators. China is the perfect example of power and dominance and dreams gone awry.

Consider this: there is a great ranting and raving about Taiwan by China's leaders. "Taiwan is a part of China," they thunder, "and cannot declare itself independent." The Chinese leaders will not stand for the Taiwanese declaring themselves free even though Taiwan is, in fact, free of any Chinese influence.

This state of affairs would be funny if it wasn't so perilous. China today has absolutely no impact on Taiwan from an economic standpoint. Indeed, the per capita wealth of Taiwan is substantially more than the per capita wealth of China. This is the result of at first a capitalist dictatorship and lately a capitalist democracy. China certainly doesn't need Taiwan. It already has the direct control of more than a billion people and cannot take good care of the vast majority of them.

Still China insists that Taiwan is part of China and will always be part of China, and if necessary, China has the right to go so far as to invade the island in order to keep Taiwan a part of this last vast dictatorship.

The Taiwanese do not want to be under the bootheel of the communists. The Chinese people do not need more people with which to share their poverty. In short, there is no reason that mainland China needs to take over Taiwan except power. The need to dominate, the predisposition to power—that is what drives the aging leaders of China. It is the insatiable thirst to dominate more and more people.

Taiwan is as rich as it is because it is a democratic nation. China is as poor as it is because it lacks democracy. The Chinese people can change this; they can force democracy on China. Twenty years ago, that would have sounded ludicrous, but today, it is not only possible, but also even remotely probable. Hungary was communist, dogmatic, and poor; then the communists were overthrown, and today democracy thrives. Poland, too, got rid of its dictatorship. Even Russia, that big bad bear of the north, is now at least democratic if not fully capitalist in makeup. Russia will be covered in the next chapter a bit more completely. Suffice to say, if the Russians can throw off the yoke of tyranny, then it is also possible for China to do so.

China is not alone in its inability to tame its leaders. Many countries are learning, albeit slowly and hesitantly, how to do just this. One of the why questions is "Why did South America and Central America and Mexico develop one way while North America became a dynamic powerhouse and the USA is today the single great power in the world?" All these countries were discovered at the same time, occupied by Europeans, and they are repositories of vast quantities of natural wealth. Why then was the development of the two geographically similar areas so different?

There are a number of differences. Most North Americans were, in the years leading up to independence, of English descent. The balance of the western hemisphere was settled primarily by Spain and Portugal. Spain and Portugal had absolute monarchies while England had developed a counterweight to the divine right of kings in their parliament. So there were differences in the mindset of the settlers at the very beginning of settlement.

The single biggest difference between North and South America is that North America has always had democracy while Mexico and South America have, until recently, been under dictatorships. There are additional reasons, which we will be getting to. But far and away, the difference between North and South America is reflected in one word—*democracy*.

But even after Mexico, Central America, and South America had thrown off their European shackles, these countries wallowed in misery and poverty. Many still do today. And the reason is dictatorships. These countries were ruled by alphas for the benefit of the alphas. Today, these countries are trying to break away from the disappointments of dictators, and they are attempting to embrace democracy. It is a long, lonely road full of potholes and pitfalls with no guarantees of success. But democracy, true democracy, is the key. The sooner South and Central America embrace the rule of the people, the better-off they will be.

Democracy is vital for any country if it is to reach its full potential. Democracy ensures that leaders behave. It ensures that decisions are made to benefit the greatest number. It ensures that unpopular individuals who behave with impropriety are removed. Democracy lays the groundwork for all that follows.

Today, three of the world's countries that are the worst off are Cuba, Laos, and North Korea. In each country, a single dominance system exists, and hundreds of thousands of people live in squalid, miserable conditions all to feed the dominance desires of some male with elevated testosterone levels. All these countries are poor, all are tightly controlled, and all this is nothing more than dominance writ large. There is also a strong delusion here. Some people, maybe even a majority, feel that to give what one can and take only what one needs is the true nature of human nature. They believe that people are anxious, even predisposed, to live this kind of unselfish, altruistic, caring, concerned, and self-sacrificing lifestyle. It is possible, perhaps even probable, that a majority of people might be conditioned to accept this way of life as a desirable goal. But there is a small, or not so small, minority that craves dominance. Then there are others who will assist this minority as long as their needs for rank within the single dominance hierarchy are met. At the bottom will be the vast majority of people with no rights, no redress, and no say in their own future. And always there is the total disrespect for human life. The true essence of the evil of alphas: life lost on an unimaginable scale.

Dictatorships demonstrate appalling interchangeability. An individual or group of individuals seize power—usually in the name of the people—and promptly go on a killing spree. Lenin and Stalin took over Russia in the name of the people then commenced the slaughter of millions of those same people. Hitler became absolute dictator over Germany and promptly began killing Germans on a massive scale. Mao Zedong became the great helmsman in China and murdered millions for real or imagined reasons. Pol Pot gained power in Cambodia and oversaw the annihilation of two million Cambodians. Castro took over Cuba and slaughtered tens of thousands of Cubans whom he perceived to be a threat to his power.

The carnage is not limited to communist dictatorships. Hitler, after all, hated communism. But the slaughter is much more common in the communist countries than it is in other types of dictatorships. What is really terrible about all this is that it truly represents the banality of evil. Well over a hundred million people were lost in just the twentieth century as a result of the behavior of a handful of men, and no one really cares. We saw no marches when Mugabe murdered. We saw no sit-ins when Pol Pot butchered millions. Mao died in bed; so did Stalin, and no one notes, no one cares. If we are so kind and altruistic, why is it that no one cares?

One could not ask for a better example of dominance than that of Cuba and Fidel Castro. Nor could one present a better example of the dire need for democracy in this poor Caribbean country. Cuba truly demonstrates the fate of countries unable to control their alphas.

Castro ruled absolutely without any redress on the part of the ruled. In 2003, he jailed seventy-five individuals who, he thought, might pose some small challenge to his dominance. He did this in spite of the widespread condemnation from countries around the world. Sweden made it clear that the jailing of these individuals could harm Cuba's relationship with the European Union. Both Canada and Italy protested, but to no avail. Castro saw the actions of these courageous men and women as a challenge to his absolute authority. They had to go in spite of the potential international harm the actions might cause. Nothing, absolutely nothing, will stand in the way of an unrestrained alpha's lust for power.

Cuba is not alone in its misery and destitution. North Korea is a basket case, and Kim Jong Il is chief basket maker. Again, the example is exquisite. At the end of World War II, Korea was divided into two separate countries. South Korea, like Taiwan, started out as a political dictatorship with a capitalist economy. North Korea started out as a communist dictatorship with a command economy. Today, South Korea is democratic, capitalist, and rich. Communist North Korea is poor, despotic, and starving. The reason, we are told, is that North Korea has a food shortage due to drought. But think about it. Korea is a peninsula; what happens in North Korea should also happen in South Korea. If North Korea has, as a result of the drought, a food shortage, then so, too, should South Korea. But South Korea has plenty of food and freedom, so perhaps it is not the drought at all.

It is not as if comparisons cannot be made, even simple comparisons. South Korea is a vibrant, well-off, free, capitalistic democracy. North Korea is a poor, starving, and bankrupt communist state. We can make other comparisons also. The United States is the richest country in the world; it is also the freest country in the world. Cuba is poor, despotic, repressed, and communist. Canada, Sweden, Switzerland, England, France, and Denmark are all rich, strong capitalist democracies. Everyone wants to get into these countries. Laos is poor, backward, and communist. Vast portions of the populations of Cuba, Laos, and North Korea want to get out of these countries. Communist countries must build fences around themselves to keep a restless and dispirited population imprisoned. Capitalist democracies must build fences around themselves to keep people out.

Do we begin to see a pattern here? Is there something population dynamics wants to tell us about the relative worth of various political systems in place around the world? Democracy is the key. Democratic countries control their alphas. Democratic countries give everyone an equal chance to achieve their full potential.

Democracy is the only way to go. If every country in the world today were a democracy, we would all be better-off. In a true democracy (unlike the People's Democratic Republic of Vietnam, which is neither democratic nor a republic), the people exercise control over their alphas. Alphas know that in order to achieve their goals, they must conform, more or less, to the will of the people.

Economically, each country must be capitalist. As we shall see, it is only capitalism that provides the greatest amount of consumable wealth for the greatest number of individuals at the least cost. That is not to say that unfettered capitalism is desirable. Reasonable rules can and must be established to restrain the capitalist visceral desire for power and dominance. Just like our political alphas, our capitalistic alphas can and will perform within established boundaries if their status depends on it.

Each country must also be secular. The influence of religion has had far more negative effect as opposed to beneficial impact on our lives. With one or two exceptions, each belief system claims for itself the exclusive right of being right. By implication, all other religions are wrong. By extension, the true believers have not only the right but also an obligation to convert the great unwashed masses. And if conversion is resisted, the true believers have the right to destroy unbelievers. We will pursue in some detail the seamy side of religion where the claim to know the will of God translates into the subjugation of man. Power and dominance writ large.

Intelligence is the hallmark of man. It truly separates us from all the other animals. It is our defining characteristic, and it has, literally, saved us from extinction many, many times. It is a powerful tool and needs to be used. Today, far too many of us cannot read, cannot write, or even do simple math. Countless cultures have no number past 10, using the local word for "many" to cover any number they cannot count on their fingers. Over half of the world's people are not educated, and many more are not educated to their full potential. It is a terrible loss for all of us when a genius lives and dies uneducated. We must educate all individuals to their fullest potential.

A country, to fulfill its obligations to man and to humanity, must be fair and just. Everyone must share the same just and fair treatment. A country cannot be known by how it treats its majority but by how it treats its minorities. This is the key. Everyone has the same rights, the same benefits, and the same chance to succeed. Everyone must be treated the same, a lesson even the United States only recently came to learn.

It begins with an assumption. Every decision, every tax, every law, every aspect of every interaction between the citizens and their government begins with an assumption. Laws are passed based on an assumption. The very fabric of government is based on an assumption. Officials are elected to public office based on an assumption on the part of the electorate. And there is nothing anywhere that validates the veracity of that assumption. Why do potential office holders seek office? Is it because they truly desire to do good for the rest of us, or do they seek office in order to seek power? We assume that they have our best interests in mind, and that assumption is wrong.

In any political system, alphas do not seek office out of a desire to serve or to do good or to benefit mankind. Alphas lead in response to a genetically influenced desire for dominance. They cannot be stopped; they can only be controlled. Potential office holders may believe that they have the best interests of their constituents in mind when they seek office. They may even feel in their heart of hearts that they can make a difference with respect to the electorate. It is delusion. It is power and dominance that they desire.

This is why democracy is so important. Under democratic systems the followers determine who their leaders will be, but even more important, the conditions under which leaders are allowed to lead. In the United States, the president is limited to two four-year terms. I am sure that Bill Clinton spent many a night attempting to figure out a way around this limit. Bill Clinton is the consummate alpha; his motivation was and is genetically influenced. He wanted to be president not to do good, not to rule wisely or well, but because his genes told him he was destined to lead.

We should not be at all shocked at Bill Clinton's behavior while in office either, especially his attitude toward women. The goal of any species is to survive and reproduce. The goal of any social species is to improve survival and procreation by establishing a dominance hierarchy. The most important male gets first crack at females in estrus or out. Bill Clinton was simply following his genes.

Democracy is the key. When followers get to choose their leaders, the leaders are forced to function within the parameters of the social group. It is the leaders who lead, but it is the followers who determine the boundaries of leader behavior, and if the social group is strong enough, it can make its leaders behave.

It is not just a case of democratically electing officials. Once leaders are in office, followers must continue to exercise control over them. One of the most fundamental delusions with which we afflict ourselves is that as a species, we are, by nature, economical. That under normal conditions, we will perform to the best of our ability, behave responsibly, and conduct ourselves ethically. Reality continues to demonstrate otherwise. But we eschew veracity for the rosy, cozy bosom of delusion. We assume that the officials we elect will weigh the needs and wishes of individuals against the needs of the nation as a whole and produce a balanced response that will maximize benefits for both while keeping detrimental results to a minimum.

We are deluded. Our elected officials, by and large, are interested only in themselves and will do whatever they can to maintain their position as elected officials. Those individuals who aspire to high governmental office do so for the power that the position offers rather than to serve the needs of their constituency.

In the United States, these people are placed in a very high management position in the single largest economic entity in the world. Very often, they have little or no experience for the job, little or no understanding of prudent economic principles, and absolutely no need to be temperate in their management of the economy. They have all the perks and benefits of the position without any of the problems. They also have the power. They are sought out and flattered by special interests with the very real possibility that some special interest of the special interests will be dealt with favorably. They are praised, complimented, courted, and fawned over. This is something to which they desperately want to become accustomed. One example and then we need to go on.

At the end of 2003, Congress passed a Medicare bill that was trumpeted as a cure for the high cost of prescription drugs. In the bill were two provisions of interest. First, the bill expressly prohibited the government from negotiating drug costs with drug companies. The one entity that was about to become the single biggest customer of the drug companies was, by law, forbidden from getting quantity discounts. The second provision expressly prohibits anyone getting federal funds from buying drugs from other countries. This is our government in action.

There is no quid pro quo, say, our elected officials, although I have as yet to see any good, sound reasoning for these provisions. It is clear that such provisions as outlined benefit the drug companies, the same drug companies that donate millions and millions of dollars to various political candidates to finance their reelection campaigns. But there is no quid pro quo.

Which brings me to the last requirement for a truly democratic country where the followers really control their leaders. Any country electing officials must insist on term limits. We must limit our leaders to one term at city level, one at county level, one at state, and one national term. The term can be as long as six or eight years, but it needs to be one term and one term only.

To make leaders behave, we must first understand why they do the things they do. Most of their actions revolve around keeping and, if possible, improving their position. They like the power, the position of authority. They like the pervasive feeling of dominance, the status of the position, and, most of all, a reserved parking spot—close in, free—at Reagan National Airport. That heady feeling of whipping into a parking spot reserved for *moi*. No endless search for a space, no fights with others over who got there first. They can just wheel in and park. Ahhh, there's nothing like it.

Our legislators take money from special interests and do favors for specials interests in order to get reelected. If the yoke of reelection money is removed, then our esteemed political leaders may just perform a little more rationally.

And here is another thing. Our senators and representatives and president run the single wealthiest country on the face of Earth. They spend trillions and trillions of dollars, much of it unwisely.

Let's get some people into office who can wisely and economically spend the people's money. One hundred and fifty grand or so a year just ain't going to cut it. If we pay our nationally elected officials a million dollars a year, we will attract the kind of management we need. We could pay them one-third of the money each year and keep the other two-thirds in reserve. This would be paid out at the end of a six- or eight-year term only if the politician has served without a whiff of scandal and without any corruption, without any offense or outrage to common sense. We would have, literally, the best politicians money could buy. But it would be the people's money, not the special interest's money.

Democracy and reasonable control of our elected officials is the cornerstone of a situation where the greatest good can be done for the greatest number. The followers must exercise their ability to influence their alphas. If they do not, the alphas will end up controlling the followers and doing not-so-nice things to ensure the retention of that control.

Democracy is the first step in the creation of a viable, prosperous free state. It is not enough in and of itself to ensure the greatest potential for the greatest number. Capitalism is another key. Even in dictatorships, the people are better-off in a capitalist economy than in a socialist economy. The reasons are simple and straightforward. They have nothing to do with kind, altruistic, benevolent behavior and everything to do with the true nature of humans.

CHAPTER 14

Why do we have economies? Why are they necessary in the first place? Our hunter-gatherer ancestors did not have economies. Each individual made all the tools, equipment, and clothes that he or she needed to survive. Little of value was owned by anyone. Perhaps they had a lucky stone or an amulet or a pretty shell, but nothing else that wasn't essential to their existence. Many Stone Age cultures today do not have economies, although contact with civilization demonstrates to these individuals the plethora of potentially superior tools available. The demand for these superior tools is the first step in the budding realm of economics.

But our hunter-gatherer ancestors had no such tools or access to any culture that had such tools. For 160,000 years, we existed with only simple stone tools, simple wooden tools—perhaps something made from bone or horn or even ivory in places, but only survival tools. Something that made it possible to stay in place, to just survive. Of course, we had fire, and it is the possession of fire that, in all probability, allowed us to survive and populate the globe.

Then as population densities gradually increased and hunting-gathering became less and less possible, we slowly and reluctantly converted to farmer-herders. But much remained the same. Each individual could more or less do all the necessary chores connected with farming and herding. Then as civilization began its first stirrings and surplus populations started to appear, we developed our first specialists.

For a long time, our farmer-herders had the benefit of a limited number of these specialists. Workers in clay, pot makers were among the first. The ability to cure and tan hides was also an early specialty, although not as specialized as pot making. As we crept from the Stone Age into the Metal Age, additional specialists sprang up. Smithies were an important and necessary component of early farming communities.

Our economies were still fairly simple, a certain-sized fired clay jar for a certain amount of grain. A metal axe for a sheep or a goat or more. Specialists had to eat but did not produce anything eatable. Farmer-herders produced food but needed specialist tools to do their work efficiently.

At about this time, looms were developed for the manufacturing of cloth. The looms could produce more cloth than individual weavers needed. The weavers in the village might even produce more cloth than the entire village needed. Now steps in a new element to the economic equation—the middleman. The trader, a product of surplus population, came into being. A trader would trade cloth for copper ingots, and copper ingots for grain. A trade route might be established. Cloth and grain were carried to the copper mines where the metal could be traded for food and clothing.

But how much is how much? How much cloth for a bar of copper? How much grain? The weaver and the farmers were not going to work for nothing; the miners and the smelters were not going to give away their metal wealth without just recompense. And that was the rub: how to determine and how to pay just compensation. It is the labor that must be rewarded. For the weavers, the wool was free, a by-product of sheep and to be exploited. The wool must be cut from the animal, carded, spun, and woven—all laborious, all tasks applied to a natural resource, the wool. The miner must mine the copper ore, which is free in the sense that it is a naturally occurring resource. The ore must be smelted and melted, and the melted metal poured into a mold. All this constitutes labor applied to natural resources. The farmer, too, must plant the seed, water the crop, and harvest the grain. So it is always the labor that must be recognized and paid for. All consumable wealth is the result of labor applied to natural resources. This is the first law of economics.

But how to compensate for all that labor? Over time, the early traders came to use coinage as the method of accounting for labor. They could have used sand, but sand is too common. They could have used water (a precious commodity in the desert), but water tends to disappear if left unattended. Copper was used in some places, but the metal was far too valuable as a tool to make it a good currency. Silver and gold eventually came to be used because both were relatively rare and of little use for anything else.

And so an economy or, more realistically, many economies were born. Trade graduated from strictly barter to a system where all parties agreed that a bit of metal, worthless in and of itself, represented a specific amount of labor applied to natural resources.

But the manufacturing of consumable goods did not constitute the total cost of the product to the consumer. One must consider the trader; he, too, has costs. There is the cost of stock and feed, of laborers to handle the stock, of accountants to track produce, and, last but not least, profit. All this had to be added to the price of the produce in order to ensure that everyone—from the sheepshearer on through the weaver, from the miner to the smelter, to the farmer right on through the trader—were adequately compensated. There was also the question of taxes, levies, and duties to be considered. All these costs together with others determine the consumer price. And the consumer, with the purchase of the item, paid all these costs. This brings us to the second law of economics: the consumer pays for everything.

The reason we have economies is human nature. The weaver could just leave piles of cloth outside the cottage door for anyone and everyone to take as they please. But shearing and carding and spinning and weaving are hard work. Calluses form—thick, tough, and crippling after a while. Making cloth isn't easy, and while the weaver is good at his trade, it is a chore, not a joy. So, too, the copper smelter. His work is dangerous and difficult. The tunnels are small and cramped. The ore is hard to extract. The smelter is hot and extremely hazardous. The farmer, too, spends long scorching hours in the field plowing, sowing, weeding, and harvesting. The herders' life, too, could be dangerous. Wild animals prowled the pastures, and herders were just another game animal to them.

All these occupations were undertaken more out of necessity than desire. The labor was constant—dawn to dusk, day after day, a lifelong litany of drudgery and despair interrupted by the occasional instance of absolute terror. A mine cave-in, a wolf among the flock. Any one of a huge number of possibilities lie in wait at a point in time when there was little in the way of law and order. It was not the best of lives, necessary if one was to survive, but certainly not to be done out of feelings of altruism or selflessness. Human nature just doesn't work that way.

We labored and mined and farmed because we had to. Game was gone or was so far away that it wouldn't pay to bring it back. We wanted the basics at first—a place to sleep, a woolen cloak, a bit of flatbread, and a jug of water. And for a long time, that was all we had—just the very basics. We became good at potting and tanning and smelting and farming and the one thousand and one things connected with survival and reproduction.

We multiplied and diversified and learned. We got to a point where some of us, the best of us, could produce more than we needed. Indeed, even more than we needed to trade for other things that we needed. And because of human nature, the surplus wasn't just given away. Because of human nature, this bounty was kept for a rainy day, a bit of bad luck, or just because the keeper wanted to do so. Surplus population begets surplus consumable wealth, and surplus consumable wealth begets economies. We were well on our way toward civilization and wealth, tokens of wealth, and, most of all, an economy that gave value to that wealth.

As we have seen, two fundamental laws govern the implementation of all economic systems. The first law states that all consumable wealth is the result of labor applied to natural resources. Consumable wealth cannot be voted into existence. It cannot be wished into existence. It cannot be legislated into existence. It only occurs when labor is applied to natural resources.

Consumable wealth is easy to define. It consists of any product used by the human animal for any purpose. The only other criteria for consumable wealth is that it has a life span. Anything that is made by applying labor to natural resources that will someday, sometime get used up, worn out, or disappear after usage is consumable wealth. The food we eat is, literally, consumable wealth. The clothes we wear, the cars we drive, the houses we live in, the toys we play with—all consumable, all with a lifetime, all consumable wealth. But there is more, the roads we drive, the bridges, the factories, the boats (or is it ships?); in short, just about everything made with the exception of art.

Art is not consumable wealth. It sometimes has a lifetime, but nominally it is eternal. If the French take care of the *Mona Lisa*, there is no reason to believe that the painting could not exist forever.

Consumable wealth must have a lifetime, be it long or short. Food has a relatively short lifetime. An orange, for instance, starts as a blossom in January here in the US Southwest. It grows for almost a year, turning orange in the December after the January in which it was born. It is picked, sorted, packed, stored, shipped, displayed, bought, and eaten—that is, it is consumed.

A car has perhaps ten to twenty years of use before it is worn out; a shirt, perhaps three or four or more years of use. A house has as much as two or three hundred years or even more of use. But all are consumable wealth and only occur as a direct result of labor applied to natural resources.

The second fundamental law of economics is that the consumer pays for everything. The fact that the consumer pays for everything is self-evident. The consumer may or may not be the laborer who produced the wealth, but when you buy a house or a shirt or a loaf of bread, you are paying for everything connected with the production of that particular piece of consumable wealth.

The loaf of bread is not just wheat—ground, mixed with water, and baked; although at one time, that labor, along with the sowing and harvesting, was the total labor investment. The farmer planted the seeds, which he paid for (and consumed by planting in the ground). And the price he paid included the cost of growing the seed; the cost of bagging the seed; the cost of transporting and storing the seed; the taxes paid by the grower of the seed; the social security taxes paid to the government on behalf of the laborers who worked on growing the seed; the cost of the equipment to plant, till, weed, and reap the seed; the fertilizer used to nourish the seed; and uncounted additional costs that go into growing wheat seed.

The farmer, too, has expenses that he must account for. He paid labor to sow, weed, and harvest the wheat; equipment to accomplish same; and taxes. He must pay real estate taxes; sales taxes; income taxes, and, in places, a use tax; even a tax on gas. A tax on grease, a tax on vehicles—all kinds of taxes, and each is added to the price of the wheat sold by the farmer.

The miller, too, has his (or her) share of costs—labor, taxes, overhead, vehicle costs, heating costs, building costs—all of which must be added to the cost of the flour. The baker, too, has costs as does the grocery store that sells the loaf of bread. In all these instances, these costs are the result of labor applied to natural resources or taxes, and in all cases, these costs must be considered in the context of determining the price of the loaf of bread.

Labor is applied to natural resources using units of time. Money could be used, but the value of money varies all over the world, and some governments have the bad habit of simply printing more money without creating the necessary wealth with which to validate the amounts printed. This dilutes the value of the money. Time is constant. It is the same minutes and hours for everyone, and so time must be the constant criteria for gauging economic systems.

In 1938, a loaf of bread costs 9¢, and the average laborer earned $1,321 a year. In 2002, a loaf of bread costs $1.47, and the average laborer earned $23,870 a year. Bread at 9¢ a loaf sounds like a real deal but only until you do the math. In 1938, the average laborer would have to work 8.5 minutes in order to earn enough money to buy a loaf of bread. In 2002, that same laborer would only have to work 7.68 minutes to earn that same loaf of bread. The 0.82-minute difference doesn't sound like much until you start to multiply it by the number of loaves of bread produced every day. Millions of man-hours saved every day by so simple an item as baked bread.

The reason that bread is less expensive in time today is because time is, in effect, the total cost of the bread. The natural resource, i.e., the land used to grow the wheat, is free. It is free in the sense in that it has always been there and did not have to be manufactured. No labor was required to bring the land into being.

Labor was required to prepare the soil, plant and tend the wheat, harvest and store the grain, and then sell it to a mill. Labor was required to mill the wheat and grind it into flour, process the flour and convert it into bread, and transport and finally sell it at the store.

Because modern technology has managed to save 0.82 minutes of labor, a loaf of bread costs less in labor time today than in 1938. The more economically labor is applied to natural resources, the greater the amount of consumable wealth there will be for every-body. Because the consumer pays for everything, it is in the con-sumer's best interest that consumable wealth be as inexpensive as possible. This applies to all consumable wealth from that as small as a loaf of bread to that as large as a car or even larger.

In the United States, the iron ore required to manufacture an automobile is mined by companies. These companies are not in business to mine iron ore; they are in business to make money. In order to make money in competition with other companies mining iron ore, they must watch carefully each minute of time expended per ton of iron ore that is shipped. The iron ore itself is free. The total cost of getting the product to the shipping point consists of labor applied to the extraction, processing, and transportation of the ore. Every technological advantage is sought; every cost-cutting measure is seriously examined to determine its suitability for implementation. To make a profit in competition with other companies, each is motivated to supply iron ore at time costs as low as possible.

And so it is, all the way down the line to the finished automobile. Transportation is competitive; the boat that is the cheapest gets the mostest. Steel mills likewise compete. They offer better quality at lower prices, driven by the need to stay competitive and make a profit. Auto manufacturers, too, compete. Ford competes with Chevrolet and with Dodge and even with imports from other countries. All this competition to reduce costs and to save time are reflected in the final costs to the consumer.

Only by knowing the total cost that goes into a product or the cost of producing a part of the product can the maker determine if the cost of the product truly represents the value of the labor that has been applied to natural resources. A laborer needs to know all the costs connected to making a product, and this only occurs in a capitalist economy.

If these costs are unknown, then the laborer has no way of knowing the true cost of the product. Consider the miller. If he continuously underestimates the cost of producing flour, he will go out of business. The miller buys a bushel of wheat from the farmer, which he then consumes by milling that wheat into flour. He then sells the flour to a baker, who consumes the flour by turning it into bread. The baker pays the miller for the flour, but because the miller failed to accurately compute costs, the baker pays less than what it costs the miller to make the flour. After meeting his expenses, the miller finds he is unable to buy any more flour from the farmer and goes out of business.

What if we just ignored all costs? What if the farmer just gave the wheat to the miller who milled it and gave the flour to the baker who baked the flour into bread and then just gave the bread away?

Sounds fine in theory—no costing, no profit motive, not even any taxes (all taxes are reflective of labor). Unfortunately, human nature precludes this from happening. We are not—many of us—the caring, kind, altruistic, unselfish, loyal, truthful, honest scouts that some think we are. Most of us will not submerge our psyche in the well of greater good out of purely selfless motives. One may find, on occasion, the random individual or even a group willing to cooperate unselfishly, but for the population as a whole, it ain't gonna happen.

Any economic system must account for all the labor applied to natural resources. Capitalism is the method whereby this is accomplished. Capitalism is a process whereby all the labor applied to natural resources is kept track of. And this is why capitalism works and socialism/communism doesn't. Those who espouse socialism/communism are operating under an economic delusion.

Consider a widget. Everyone wants a widget. Under the capitalist system, 800 laborers can produce 1,000 widgets in one hour. Under the socialist/communist system, it takes 1,200 men to produce 1,000 widgets in one hour. The whys of this are simple enough. Under the capitalist schema, every minute of labor is looked at and analyzed and evaluated to see if somehow in some way it can be eliminated or reduced. The thrust of the capitalist process is to produce an economical widget.

But under the socialist/communist system, the plant manager has an apathetic son-in-law who is vice president in charge of super-duper pooper scoopers and does little to advance the economical production of widgets, but whose labor still counts. The superintendent of the plant has a large extended family, most of whom work at the plant not always economically, but whose labor is still counted. The many foremen (twice what is needed) also have their friends and relatives who are employed at the factory.

So what? It takes longer, so what? Both plants produce widgets; the job gets done. And besides, we have loftier things to think about.

But everyone wants a widget. Everyone must have a widget. It is essential, it is mandatory, and that is the rub. Under the capitalist system, 800 workers produce 1,000 widgets in an hour. This is accomplished by applying labor to natural resources. Under the socialist program, it takes 1,200 workers to produce 1,000 widgets in one hour. Since everyone wants a widget, in the capitalist camp, we have a widget for each worker and 200 left over. In the socialist camp, we have only 1,000 widgets to satisfy 1,200 laborers. Capitalism routinely produces excess consumable wealth, and socialism routinely produces shortages.

The two-hundred-widget shortage under the socialist/communist economic system is only the beginning of the problem for the laborer. Distribution is also a critical factor in socialist economies, and here, human nature, again, rears its ugly head.

The manager of the widget plant must, of course, have a widget—and a widget for each member of his family. Those widgets come right off the top. Not only that, but the plant manager also expropriates extra widgets to curry favor with other powerful widget wanters. For the party bosses, for extra rations, for luxuries not afforded the masses—all this will require extra widgets. So, too, the superintendent and the foreman—they all require extra widgets for various and sundry reasons, all taking from the communal pot. And in the end, there is only 300 or 400 widgets to satisfy the needs of the 1,200 socialist workers who actually made the widgets. This is how human nature works.

We see it in every socialist country in the world—shortages of food, shortages of housing, and shortages of wearing apparel. The list goes on and on, and that is why socialism doesn't work.

You will notice that nowhere was money mentioned in the widget drama. It is not important if the laborers are paid $1 per hour or $10 or $100 per hour. Since the consumer pays for everything and since all costs consist of labor costs applied to natural resources, it becomes immaterial what the per hour cost of the labor is. The key is the amount of labor applied to the natural resources to produce the final product. Where it is controlled and minimized, economies prosper. Where no consideration is taken of labor, we have shortages.

The French unions had an idea a while back that they thought would go a long way toward improving the lot of the French worker. The worker would get paid for forty hours of work but would actually work only thirty-five hours. I cannot imagine what these officials were thinking, but analysis of the idea quickly demonstrates the futility of the concept. The laborers are paid for forty hours but only work thirty-five hours. As a result, one-eighth of the cost of manufacturing the item must be added to its cost. When the laborer buys the item, the cost of the item reflects this one-eighth increase, so the laborer's money only goes seven-eighths as far, and in effect, the laborer is being paid for thirty-five hours of work.

Carry the idea to its logical extension. The laborer will work twenty hours but will get paid for forty hours. Because all consumable wealth is the result of labor applied to natural resources, the item will cost twice as much as it would otherwise. The laborer, in buying the item, will pay twice as much money, effectively reducing his income to the value of twenty hours per week.

We do not need to go beyond our own US boundaries to see idiocy presented as enlightened thinking and how perception colors our delusions. Sometime back, Michael Moore made a movie. The movie was called *Roger & Me*. It purported to show Moore trying desperately to find the CEO of General Motors. Moore had questions. He wanted to know why GM had packed its bags and left Flint, Michigan, for sunny southern climes. Moore was not successful in his quest.

Then again, he didn't want to be. The whole point of the movie was to paint the unemployed workers at GM's Flint plant as victims and the CEO of GM as a villain for leaving them jobless. Moore, of course, knew the answer to why GM moved but did not let that stand in the way of his presentation of the Flint autoworkers as being unfairly and unjustly denied the right to work by an evil, uncaring cabal.

For years after World War II, the United States had a virtual lock on automobile production. We were the only large industrialized country in the world that had not been damaged or destroyed by the war. We also had a huge labor force, the direct result of mobilization for war. A labor force with lots of money to spend, money that couldn't be spent during the war because of rationing. The car business boomed and boomed and boomed.

Labor unions formed for the very real and very good purpose of protecting workers' rights. Capitalists are not known for their generous, sharing nature and so must be watched closely. But so must unions. For years, the United Auto Workers Union milked the situation for all it was worth, getting raise after raise, concession after concession, and benefit after benefit. The big three automakers didn't like it but didn't contest the greed either. All three paid the same; all three passed the increased costs on to the car-buying public. Cars quickly became relatively expensive compared to many other items of value. But the buying public had no alternative.

Then came Japan. It took years. In the beginning, Japanese goods were deemed inexpensive although of poor quality. But the perception changed over time, and Japanese goods were soon seen as inexpensive and of moderate-to-good quality. Then came the cars. Japanese imports were equal to or better than American made and, in many instances, at half the cost. Japan quickly took a huge bite out of the American car market.

The big three tried every trick they could think of to boost lagging sales. They cut profit to the bone, they tried patriotism, and they even asked for union concessions to meet the prices set by the Japanese imports. Even with minimum profit, the American cars were more expensive, and many buyers were more loyal to their pocketbooks than to the car companies. But it was the unions that really cost the carmakers. As we have seen, all consumable wealth is the result of labor applied to natural resources. It was union labor costs that were keeping American automobile costs uncompetitive, and the union meant to keep it that way.

By the time the problem really became a matter of survival for the auto manufacturers, a union employee was making $30 an hour for putting nuts on bolts. Union rules made it next to impossible for a front-seat installer to install a rear seat. It kept a wheelman from touching a brake pedal or a fender man from installing a hub-cap. Union rules were catered to for years because the companies could pass along to the car-buying public the costs generated by these rules. Over time, these costs became onerous. Benefits, over-time rules, and other deferred compensation drove costs through the ceiling. Something had to be done.

Think about it. GM had land, buildings, equipment, tools, and track worth billions of dollars in Flint, Michigan. GM had an experienced, knowledgeable workforce in Flint, Michigan. GM had a huge investment in staff and supplier capability in Flint, Michigan. GM would not throw all that wealth and capability away lightly. There had to be some good, sound economic reason for GM to leave all that behind and move out of Michigan.

And there was. GM could stay and go broke because the cost of its vehicles was too high, or it could move to a place where labor costs were more reasonable. It could move to a place where work rules would allow the introduction of labor-saving devices and where benefit costs were acceptable.

And that is exactly what GM did. The result for the Flint work-ers would have been the same regardless of GM's decision. Had GM stayed, the company would have gone out of business, and the Flint employees would have been without jobs. As it was, GM left, stayed in business, and employed thousands of workers in other states while the Flint employees lost their jobs.

Moore did not want to find the CEO of General Motors. Had he done so, he would have been told that it was the Flint employees and the United Auto Workers who were ultimately responsible for the GM decision to quit Flint. That is not the answer that Moore wanted.

There is a strong predisposition in many countries toward socialism as a way to prevent what happened in Flint, Michigan. Uncaring, unscrupulous capitalists would be replaced by a caring, concerned government-controlled business. But cars, like widgets, are not built by caring. They are built by applying labor to natural resources, something the Flint workers never came to understand.

This is why capitalism works and socialism doesn't. It is absolutely mandatory that all labor costs be kept track of and included in the final price of the item. Capitalism does this; it is simply the process of recording labor as it is applied to natural resources. Socialism does not do this, and the inevitable result is shortage.

No country reflects this condition better than Cuba. When the communist dictator took over Cuba, things were pretty good. Castro nationalized everything, and for almost two years, the revolution rolled along pretty well—then the wheels fell off.

Cuba, under Batista, was a dictatorship but only politically. Its economy was run by a small oligarchy on a for-profit basis. These individuals had accumulated large amounts of wealth as a result. In two years, the new dictatorship went through this wealth and came up against the reality of economics. Shortages began to show up, and rationing was introduced. Rationing occurs when there is a shortage and works by ensuring that everybody gets the same small amount. With the exception of the first two years of the revolution, Castro's Cuba has always had rationing.

Today, Cubans live on 3 kilos of rice (approximately 6.5 pounds), 3 kilos of sugar, 1/2 kilo of beans, 8 eggs, and 1 bar of soap per person per month. A few other observations: Cubans can buy TVs and VCRs and even designer clothing but only with dollars, that filthy currency of the Yankee dominators. Accountants make $10 a month. A maintenance worker makes $8 a month and would have to work 4 days just to buy a single roll of toilet paper if he could find it. This is how the average Cuban lives.

Not so the elite. Fidel Castro has never owned a ration book in his life. Actually, that is not so. The truth of the matter is that Castro did, indeed, have a ration book. But it is a unique one-of-a-kind ration book. Castro's ration book was, first of all, solid gold. This is only the beginning. The ration Castro was entitled to was as follows: anything he wanted, anytime he wanted, and any way he wanted.

Let us say, Castro wanted barbecued lamb. He opened his ration book, and lo and behold, it says Castro was entitled to all the barbecued lamb he wanted. Ice cream? No problem. Open the book, and it says all the ice cream he wanted. No matter what Castro wanted, his ration book said he's entitled. If Castro wanted pork, Castro got pork. If Castro wanted French silk sheets, Castro got French silk sheets. The dominant elite get whatever they want at any time, and they do so because of human nature. Castro was an alpha, a dominant, and dominants always get the best of everything even when there is not enough to go around. All dictators act the same way. Shortages are for the peasants.

There are so many horror stories about the problems with a command economy that it is hard to know where to begin. There is this story told out of Russia. I don't know if it is true or not, but it is typical. After World War II, Russia began a huge building program. With wealth looted from the countries under its control, Russia undertook not only the rebuilding of its war-devastated manufacturing sector, but also its residential housing needs. Huge blocks of apartments were built, twenty and thirty stories of flats. All were the same—small, cheap, and nondescript. Hundreds of thousands of units were built; unfortunately, millions of units were needed. For the reasons outlined, supply never caught up with demand, and eight to ten people often were crammed into apartments meant to house a maximum of four, and that under crowded conditions.

The building began. Among the thousands of construction items needed was soil pipe. Now soil pipe is just that—pipe for the transportation of human waste. The demand was great; the buildings must go up. And to ensure that demand was addressed, the ministry of soil pipe was set up. The politburo got together and decided how much soil pipe was needed for the five-year plan. The decision was made that x number of feet of soil pipe was needed, and this figure was communicated to the ministry of soil pipe, who would then produce the required amount.

Soil pipe, at that time, was made of cast iron. Cast in lengths of up to twenty feet, it had a bell housing on one end. This allowed lengths to be fitted together and sealed with oakum and molten lead.

The soil pipe ministry calculated that it could best meet demand by casting the pipe with thin walls. This accomplished two goals. First, it made use of scarce resources—in this case, castable iron—and it also meant that quotas could be met without too much exertion. But thin walls meant weaker pipe, and soon pipe was getting to the jobsite with as much as 25 percent breakage. True, some of this was caused by the poor roads and even worse rail lines the pipe had to travel in order to get to the place where it was to be used. But the builders were not happy. Complaints were registered, but the ministry of soil pipe pointed out that when the pipe left the factory, it was in good condition. And what happened to it after that was the responsibility of the transportation ministry. It was tactfully pointed out that thin-walled pipe not only got damaged in transit, but also in handling on the job and even after installation. A point at which much time and effort was spent replacing it. But the ministry of soil pipe would have none of it. Transportation, building, and replacement problems were not its concern—only the manufacturing of pipe, and they could see no reason for change.

This went on for several years. Building construction was held up by a lack of many things, among them soil pipe. It came a time when the politburo looked at the building situation and decided it was going too slow and needed to be sped up. The problem of the soil pipe came up, and a solution was offered. No longer would pipe production be measured by the foot. Instead, production would be based on weight.

Now the whole emphasis shifted, and thick-walled pipe became the norm. Walls were now thicker than needed but cast, anyway, because each additional pound of metal cast into the pipe meant the quota could be sooner reached. Bells were lengthened and enlarged, but mostly it was wall thickness that changed. A four-inch pipe would be cast with a seven-eighth-inch wall, leaving only two and one-fourth inches of opening. Using this method, the ministry produced much less footage of pipe, but at least, the pipe worked and did not break as the old pipe did. The problem was solved, but the building industry still suffered from a lack of pipe.

There is an interesting footnote to the soil pipe story. In the 1960s and '70s, plastic soil pipe came into use in the West. Today, virtually all soil pipe is plastic, but in Russia, for years the ministry of soil pipe resisted this trend. Had the ministry gone to plastic using the tonnage figures mandated by the powers that be, they would have had to increase production by an almost unimaginable amount. New methods, new raw materials, new factories, but the same old quota. No, it was much easier to just keep on casting iron pipe.

There is another story out of Russia, and it is undoubtedly true. Sudogda is an industrial town one hundred miles or so east of Moscow. Sudogda has central heating not by individual building but one heating system for all fifteen thousand residents of the town. Because of problems with the heating systems, more than half of Sudogda's apartments have no heat. In a country where the temperature rarely rises above freezing from November through March, this is no small problem. One pensioner bought and installed an electric heater to keep her flat warm. When she got her first electric bill of the winter, it was for the equivalent of US $25, and her monthly pension was only US $26. It isn't easy being a Russian pensioner.

Sudogda's heating system was based on the use of hot water. It was poorly designed and poorly built and poorly run—typical of a command economy. As a result of these problems, by the time the water reached the outlying apartments, it had lost all its heat.

The operators of the heating system for Sudogda made sure that their flats, usually large and spacious, were always supplied with heat. While it might take two or three months for the heating people to address the complaints of the masses, a broken line or clogged radiator in the home of the elite was immediately attended to.

This is a small thing, except to the people involved who faced the possibility of freezing to death. But it is indicative of the problems with a command economy. Had Sudogda's utilities been put in private hands, all would have gotten the attention they deserved; after all, this is the only way utility companies stay in business.

Command economies cannot supply the demands of consumers because they do not consider the needs of consumers when making plans for the demand economy. It is the quota that governs, not the need. It is the plan that counts, not the wants. Be it clothes, food, footwear, furniture, anything and everything consumable, the command economy fails because it is incapable of responding to consumer demands. Instead, it is the reflection of the ideas and concepts of the single dominance system in place. Those who make the plans have what they want and what they need, even a surplus of supply. They do not care or feel the need to care about the needs of the masses. Socialism doesn't work because of the nature of human nature. The contention that socialism can work is a delusion.

*****

One of the biggest puzzles facing the communists today is not the failure of every socialist experiment ever entered into, but the failure of capitalism to collapse. This is a prime prediction of Marx: capitalism would collapse, socialism would take its place, and labor would control the means of production. Didn't happen and won't happen.

When Marx made his prediction, everything he saw convinced him that this was the only viable outcome for the economic situation of the times. There was a huge disparity between the capitalist and the laborer. Vast ostentatious flaunting of wealth by the capitalist against a constant ongoing battle by the laborer just to get enough food to eat and a bit of shelter over his and his family's head. Not a single extra penny, just enough to buy the basics while the capitalist kept all the wealth for himself. Marx reasoned that the laborer would only put up with this for a short period of time before revolting and seizing the means of production. This would usher in socialism and rule by labor or, in that most unfortunate choice of words, a "dictatorship of the proletariat."

But it didn't happen. What did occur never occurred to Marx. Labor organized and began to exercise its clout to improve working conditions. Far from wanting to take over the means of production, labor only wanted a larger share of the profits. Labor organized, and to maintain their position, capitalists realized they had to negotiate with labor in order to achieve their goals. It wasn't easy, and it took a long time, but eventually labor and capital reached an agreement whereby labor would get a wage greater than that needed to keep body and soul together. Capitalism would control the means of production.

The concept had unexpected benefits for both capitalism and labor. Excess wealth in the pocket of labor opened up additional markets for capitalists. For years, the automobile was a rich man's toy. It was very, very expensive, handmade for the most part, with a team of men building one from the ground up. They would saw and file and fit, putting each car together as an individual item—unique and only as good as the capability of the team.

Henry Ford came along at just the right time. Labor was accumulating excess wealth, and Henry Ford saw a way to make money by making an inexpensive auto—a car the average wage earner could afford. He did this by wringing every excess minute of labor out of automobile production. No more cutting and filing and fitting. No more teams of individuals building from the ground up. It was to be one man, one nut, one wheel, one fender, one door, one running board, and the conveyor belt, where frame met part after part after part until there was a complete auto.

But it took lots of people to make these cars, and by paying them a good wage, Henry Ford developed a potential market for his product. But it gets better. Cars need roads; roads require cement and asphalt and stone and bridges and sidewalks and storm sewers and all kinds of additional needs that can only be met by applying labor to natural resources. Labor that also produces excess wealth that becomes a source of additional capitalist potential. The list goes on and on—gas stations, garages, auto accessories, replacement parts, hotels and motels, restaurants, and all sorts of spin-offs necessary to keep a car running and to provide a place to be when the traveler goes somewhere.

And all because one man wanted to make cars. Not to make roads mandatory, not to supply parts for autos, or to make places for travelers to go, but to make money. Henry Ford did not put America on wheels to benefit the American public. He did it to gain wealth because we all agree that the accumulation of wealth is a desirable goal and people with wealth are looked up to, admired, and given power and dominance.

*****

Capitalist democracies face problems of a different sort. For instance, got milk? The US government had. As of 2002, it had in storage 1.28 billion pounds of powdered milk. That's billions with a *b*. This is all because the United States has the capacity, under capitalism, to produce all the milk this country needs and a great deal more besides—billions and billions of gallons of milk. When the price falls below $9.90 a hundred weight, the government is obliged to buy enough milk product to bring the price up to $9.90 a hundred weight. As a result, the government is the custodian of over a billion pounds of dried milk as well as vast quantities of butter and cheese.

Nobody likes capitalism. That statement is not, strictly speaking, true. Most of us, at least in capitalist societies, don't think much about the system at all. But among the leftist elite and the anointed, capitalism is a four-letter word. The profit motive is derided, greed is deemed inhumane, the supposed exploitation of the workers by capitalist money-grubbers, the pollution of the atmosphere, the ravaging of the widow and orphan—all the direct result of capitalist pressure, all traced back to capitalist greed, to a desire to get even richer. The capitalist need to garner ever more wealth.

There is nothing wrong with greed. Greed gets things done and economically too. Personal accumulation of wealth does wonders for the economy. It generates more disposable wealth for more individuals to accumulate, and it can be controlled. Indeed, it is controlled by rules and regulations and taxes and directives. There is not a thing wrong with greed.

But to the delusionist, greed is terribly wrong, for it demonstrates a human condition that does not square with their concept of the true nature of human nature. Humans are naturally caring and giving, not grasping and taking. We are altruistic, not selfish, cooperative rather than antagonistic, and, in general, too good for greed.

That is why, the elitists say, capitalism is bad. The profit motive sucks, and the leftist elite wants to do away with every capitalist economic system. Capitalism is to be replaced by the government ownership of the means of production—in other words, socialism. We will do away with the profit motive, turn the means of production over to the workers, whom, of course, the elitists will direct, but without a profit motive.

Economic idiots are not restricted to leftist elitists. In the 2000 election, the Greens came out of the woodwork and attempted to present the public with a wish list of absurd ideas. The following represents how misconception, aberration, and self-deception combined to demonstrate the delusion of man.

These zany insaneys have the idea that if one just holds one's mouth right and puts a good English spin on the ball, we can establish for all eternity the equality of man. Among their proposals, the workweek would be cut to thirty hours, but workers would get paid for forty. Each family would have a guaranteed income of $26,000 per year, earned or not. The minimum wage would be $12.50 per hour (this works out to $26,000 per year), and no one would be allowed to earn more than ten times that amount.

It just will not work. As we have seen, if the workers only work thirty hours and get paid for forty, the cost of goods goes up 33 percent. The price paid for goods is 33 percent more, and the job holders are, in effect, getting paid for only thirty hours of work. Nor will the $26,000 per year work for the same reason. It is difficult to understand how otherwise intelligent men and women can seriously propose such nonsense. But they do and, in the process, demonstrate two things: a total lack of understanding of the economic process and a penchant for grandiose delusion.

Karl Marx made two fundamental errors in his quest for utopia through communism. His first mistake was in making the assumption that the human species would welcome and embrace his concept of equality. He believed that the human animal could adopt the way of the ant and toil away to preserve the whole without concern for self. He believed that we would gladly embrace the concept of selflessness, that we would do good for the greater whole and willingly give up individuality.

Marx did not realize that the essence of communism in and of itself results in a power vacuum. The whole concept of doing one's best without direction or coercion eliminates any type of dominance structure. Indeed, true communism is identified as, in part, the withering away of the state. This implies a lack of direction and control from above. Marx assumed that this was the true nature of man, that he was selfless, altruistic, kind, gentle, and highly motivated to do what's best for the group as a whole. In this, he erred substantially.

Just as nature abhors a vacuum, man abhors a power vacuum, and someone, somehow, in some way will always find a way to fill the power vacuum. Such a person will do whatever is necessary, whatever is needed to maintain that power, which brings us to the second error of Marx.

The second error is one of semantics and is responsible for the deaths of more than a hundred million people. Four simple little words coupled with man's inherent propensity resulted in Marx's biggest and, far and away, the worst mistake any man has ever made.

"Dictatorship of the proletariat"—four words and over a hundred million dead. Marx could have used *control* or *rule* or *management* or *guidance* or *direction* or *jurisdiction* or *supervision* or even *leadership*. But no, he had to use *dictatorship*. The alphas, the dominants, pounced upon these words (especially the first) to establish their natural right to dictatorship. "See, Marx says so right here—dictatorship." So absolute, unfettered, unrestrained dictatorship was and is the order of the day for all communist countries. This is the one driving force for all communist dictatorships. It encompasses complete and unqualified dominance of everything and everybody.

This is accomplished by slaughtering anyone who is perceived to be a threat. Notice I did not say *was* a threat, but anyone perceived to be a threat. Whether or not the group or individual is a threat is less important than the perception in the eye of the dictator. All must be killed or rendered harmless—and families, too—because such a response will keep future restless individuals and groups from doing anything to endanger the dictatorship.

And so the games begin. The dictator may even start out with good intentions, but this is not probable. The dictator ends up spending time ferreting out plots, killing plotters, and doing other nasty things to remind everyone of the rights and prerogatives of the dictator. Kill a few disruptive elements and then order the building of some great monument or edifice or factory to be named after the great fraternal, beneficent dictator. The aim of the dictator is to retain power, to enforce dominance, and all other aspects of rule become secondary.

It is a dismal reflection on the nature of man that we behave this way. And yet one cannot argue with history. And the death of men at the hands of other men is the history of man—war, to be sure—but death in war is only a small part of the total. Far and away, the slaughter of the innocent, the helpless, the defenseless is the prime sign of man's relationship to man. It happens over and over again, and we studiously ignore the butchery of men by their fellow man. We deny the cruelty, blinding ourselves to reality, adrift in a delusion.

*****

Capitalism works because it takes advantage of human nature. The alphas, the dominants, see the accumulation of wealth as status enhancing. As we have seen, Henry Ford did not apply the assembly line (a concept far older than him) to automobiles to put the world on wheels. Nor did he do it to make traveling easier or to provide a way for the public to consume gasoline. He did it to get rich. In the process, he created a whole new industry. Not only cars but also roads and diners and motels and garages and gas stations and, yes, even junkyards. A single man with an idea and a bit of capital. It is capitalism that supplies the needs of man.

Socialism fails as an economic system because it falls prey to human nature. The practitioners of socialism do not have the same motivation as capitalist alphas. The goal of the socialist potential dominant is to kiss the ass of their superiors and kick the ass of their subordinates. The economical production of consumable wealth is far down on their list of priorities. They fail to consider the fact that to produce enough consumable wealth to satisfy the consumer of today, the economical application of labor to natural resources is essential.

It is spring of 2004, and, we are told, an economic upswing is underway. All the signs are to the good. The stock market is up, production is up, confidence is up, and even consumer spending is up—all good signs and an indication that labor is being applied to natural resource very economically.

Perhaps too economically. One of our most cherished economic indicators is not doing so well. Joblessness refuses to decline. We are producing more consumable wealth, but we are using less labor. Businesses under the auspices of capitalism have found ways of producing just as much consumable wealth but are doing it with less labor. Robots, new designs, and labor-saving devices all contribute to this state. In the long run, the trend is actually desirable. The more economically labor is applied to natural resources, the greater amount of consumable wealth will be available for everyone. The more there is, the less expensive it becomes and the more available it becomes. And at the root of this abundance is that old economic dictum: the more we make, the more I make. Capitalism in action.

But what about the unemployed? This is a problem but only in the short run. The more consumable wealth available, the more the demand. New houses need new furniture and new appliances, new landscaping and lawn equipment—the list is endless and ongoing. It ripples outward to collide with ripples from other demands caused by the consumption of wealth. It morphs into a larger and larger economic complex predicated on the purchase of a house.

The unemployed will get theirs, but it will not be easy. It is the simple jobs, the repetitious jobs that get capitalism's attention first. It is these jobs that offer the capitalists the greatest savings. So the simple jobs disappear, and it is only the skilled jobs that remain. Not doing the job but making the tools required to do the job with less labor. The call is not for brawn but for brains. And that brings us to the next requirement for a healthy, prosperous, productive state. It is time to tackle education.

Chapter 15

If we are to survive as a species, it will only occur through education. As we have seen, we hitched our future to intelligence hundreds of thousands of years ago. It was our large brain that stood between us and extinction. Our first challenge was fire. Many a time the fire was lost, and the clan was lost. Many a time the fire was lost and only regained through a fortunate accident or by the raiding of a neighbor. Over time, experience and intelligence came together, and we mastered fire. For a long time, we used it and cared for it and protected it and did everything we could to keep it from going out. We learned to keep fire.

Then at some point in our past, a toolmaker at his task struck a stone with another stone, and just as had happened many times before, sparks flew. But this time, it was different. One of the sparks flew into a small pile of punky wood one of the children had been playing with. While the toolmaker continued his task, the spark just happened to land next to a tiny bit of very dry punk, and it started smoldering. The toolmaker looked up from his work in surprise to see a wisp of smoke rise up out of the punk. Soon there was more smoke and the smell of burning wood. The toolmaker picked up the little pile and, just like he would a glowing twig, blew on it. Almost immediately, a red glow appeared; the punk popped and snapped. Another gentle breath, and a red-yellow flame appeared.

Now our toolmaker was not just another flint knapper. He was an excellent toolmaker, a fine craftsman who could see the finished tool inside the stone even before striking the first blow. The toolmaker knew he had created the fire in the punk. He knew that the sparks from the knapping were hot because they had often landed on his bare arms. He then made the connection; where there are sparks, there is fire. We had learned to make fire.

If a clan is careful enough, they do not need to know how to make fire, only how to keep it. But knowledge of the way to make fire is much like an insurance policy; in the long run, it is better to have one. It is the knowing that is important, the knowledge that made survival just a little bit easier, a bit less stressful. We survived on knowledge.

And so it was. At first we came out of the water and went back to the trees. Characteristics of our ancestors point strongly in that direction. We didn't need anything else. The trees offered protection, and we survived with little brains only slightly changed and with a bit more intellect. We lost our muzzle and our large canines, but we gained a differently shaped brain. A shape that allowed some areas of the brain to develop more while other areas developed less. It was entirely an accident, a coincidence, a chance happening, but the area most affected by the shape change—the frontal lobes—were the seat of knowledge and learning and remembering experiences. From then on, learning played a huge part in our survival.

Our first tools were pointy sticks and wooden clubs and small rocks and horn cores. We learned how to acquire, shape, sharpen, and use these tools to procure our food. We advanced to atlatls to add distance to the pointy sticks, a bow to extend the range of smaller pointy sticks, slings to throw rocks, nets, baskets, and a hundred and one different tools and tricks to survive. And so it is, still, today.

As we, in our past, were dependent upon learned behavior to survive, so, too, will our future depend upon learning. Every individual in the world must be educated to the fullest extent to which each individual is capable of learning. Learning is the real key to our survival. It is going to be a long, hard slog. It will take lots of work, dedication, and perseverance, but it can be done. It has to be done if we are to survive.

Let us go back to Africa. There is an AIDS epidemic of massive proportions in Africa. There is also the very real possibility that a young African living in some small rural village may have both the brains and the wit to conquer AIDS but only if he or she is educated. It is almost certain that all the necessary intellectual potential needed to cure all of Africa's ills lies within the population of the continent. It simply needs to be exploited.

There is a story out of South America that may or may not be true, but in any case, it does illustrate the necessity of educating everyone. Gerald Durell was a collector of wild animals and a raconteur of some of the best animal stories in the world. He tells a tale of the birdman in one of his books. Deep in the rain forest of South America lived a poor native with a penchant for illustrating birds. Durell saw his work and was impressed, impressed to the point where he tried to get drawing supplies to the man, pencils and paper being difficult to find in the middle of the jungle. The story more or less ends there but for the nagging notion that the world may have lost an Audubon to a lack of education. Think of it, someone to do for South America's birds what was done for the birds of North America lost because the native talent was not honed and directed by an education. Such a situation is a loss for all of us.

How many scientists, engineers, doctors, businessmen have we lost because of a lack of education? Not just a lack of education but a deliberate attempt to keep the masses uneducated. The better to maintain control over them.

It is the whole world that loses when people go uneducated. It is the whole world that is deprived of an individual capable of curing disease and lacking only the education to achieve the goal. We are deprived of engineers, scientists, physicists, and a myriad of other individuals who can make a solid, substantial contribution to mankind. This is why we need to educate everybody. This is why we need education.

The United States has one of the best records in the world with respect to education. The United States also has some of the finest institutions of learning in the world. We have this capability because we create so much wealth through a combination of democracy and capitalism. This is the key. It is only wealthy democratic countries that educate to the greatest extent possible. But even in the United States, the system is less than perfect, and the disparity is cultural in nature.

The story is huge. "Local Girl Makes Good" was the headline in the middle of the front page of the news in our hometown paper (circulation in the hundreds of thousands) and a full page inside complete with five photographs. The girl has graduated from high school and is now preparing to go to college. She also works part-time at a local store, tutors others who need schooling, and helps her family financially. A common situation, you might say. Many people, both boys and girls, graduate high school then go on to a higher education. I have a niece who is doing exactly the same thing, as have many people. Education is the only way to go.

But this girl is different; she is not Anglo but Hispanic, and that difference warrants a full spread in the newspaper. What is common for Anglos is rare for Hispanics, and Maria (for that is her name) is the first person in her family to graduate high school. Even a sister and a brother of Maria have dropped out of school. So Maria is doing a truly different, a truly daring, thing.

Hispanic culture decrees that females have babies—the sooner, the better. A female putting off procreation for an education is flying in the face of tradition. But the problem goes deeper. There is among many Hispanic males not only a dislike of education, but also an absolute revulsion for learning. It is deemed an honor to be unlettered. Those who show signs of intelligence are considered to be dweebs and geeks. A macho Hispanic male is tough, proud, loyal, and dumb. It's a cultural thing.

Yes, there are exceptions. There are Hispanic lawyers, doctors, and engineers. But the vast majority of Hispanic males do not see any advantage in learning, and so Hispanic culture has a high drop-out rate, a largely manual labor workforce, and an income level barely above the poverty line.

This is not just a problem in the United States. Mexico, Central, and South America have a completely Hispanic culture and a record of little or no education for the great mass of people at the bottom of the economic ladder. Much of the blame for this can be laid at the foot of the Catholic Church, which prefers its parishioners—for the most part—to be ignorant, gullible, and superstitious. We will pursue the role of the church in the delusion of man in more detail a bit further on. Suffice to say, they bear much of the responsibility for the present situation in Central and South America.

It is not just Hispanics who have a low education expectation. African Americans, too, prefer the ease of the uneducated. An education is hard work. For most of us, it is a difficult slogging grind with the added problem of not seeing any real benefit to "learning all this stuff" at the time. It is not easy, and every year, there is more and more to learn. As knowledge piles up and there is more information to pass on, education becomes more complex. When I grew up, computers were only an idea in someone's head. Today, it is essential for everyone, especially students, to be computer literate. But being computer literate means, among other things, knowing how to type. As a result, a skill once needed only by secretaries and mystery writers now becomes mandatory for all.

By looking at different cultures, we can see different approaches to education. To the leftist liberal elite, cultural judging is politically incorrect, but it is also necessary if we are to survive as a species. There are a number of cultures that esteem education and recognize the importance of giving everyone the best possible access to information and learning. The Japanese begin the process of education even before the child starts school. They do so by paying teachers to instruct three-, four-, and five-year-olds to give them a leg up when they start school. But the regimen does not end there. A child is subjected to night school, special tutors, summer school—in short, anything that might help the child learn. But what really motivates the students is the high expectations manifested by the parents, teachers, and peers toward achieving status as a student.

Sometimes, the pressure is too great, and as a result, suicide among Japanese teenagers is common. Japanese culture clearly demonstrates the potency of a cultural attachment to education. The Japanese are among the best educated people in the world. It's a cultural thing.

But the Japanese are not alone in their vigorous pursuit of knowledge. Most Southeast Asian cultures are knowledge dominated. Indeed, some Pacific coast colleges have had to limit the number of Asian students because the demand among these people for learning is so high.

There is yet another group that reveres learning, and they are the Jews. Jews are, as a cultural group, very, very successful. Some would even argue that there is a great worldwide secret plot run by the Jews, financed by the Jews, and controlled by the Jews to take over the world. The truth is much simpler and even a bit mundane: Jews love learning.

To be a rabbi is among the greatest titles a Jew can achieve, but the word *rabbi* simply means teacher, one who is authorized to decide questions of Jewish law. In short, an educated, intelligent, and knowledgeable person. To be able to discuss the book in learned terms, to be able to quote chapter and verse to make a point, to be able to explain a passage according to one or another of the ancient rabbis to prove or disprove a point, to show a mastery of the Old Testament—those are the tasks of a rabbi; and holding such a position is to be at the pinnacle of Jewish culture. To be literate, to be thoughtful, to be able to discourse in learned terms—this is the essence of Jewish social structure, and these same characteristics make possible the representation of Jews in the professions in numbers all out of proportion to their actual presence. Doctors, lawyers, engineers, business leaders, and, last but not least, teachers in vast numbers, and all because Jews love to learn.

There is yet another facet to the proliferation of professional Jews, and that is the Jewish mother. Never ever underestimate the power and potential of the Jewish mother. "Be a doctor, how about a lawyer, go, do, anything—but important." And so Jewish children are coerced, coddled, and cajoled into being "something—but important."

But Jewish mothers have an ulterior motive. There is a delicious irony to the coddling and cajoling by the Jewish mother and her concern for her children's career. Jewish mothers may think that what they do is best for their offspring, but their real motive is what is best for themselves. The height of bliss for a Jewish mother is to be able to brag about her children when talking to other Jewish mothers. "My son, the doctor" is a much beloved phrase and is spoken with pride by Jewish mothers wishing to turn coffee mates green with envy. This is also the reason we are blessed with Jewish doctors in numbers all out of proportion to the actual population of Jews in the United States.

It is simple; it is straightforward. Cultures are dominance systems, and as the culture believes, so do the culturalists. If a culture cherishes and extends status to learned individuals, then that culture will be educated and knowledgeable. Cultures that eschew education will be ignorant, gullible, and superstitious.

But education in and of itself is not enough. What is taught is also important. As we have seen, creationism is a patently false absurdity. And yet a sizable portion of the public wants it taught in the public schools. Creationism can tell us nothing about man, nothing about how and why we are what we are, nor can it offer any reasonable solution to the problems that plague mankind.

Evolution, on the other hand, presents us with a possible explanation for human behavior and potential ways to understand and control that behavior. We are evolved animals. The statement may insult some, but it offers an insight into why we do the things we do. Evolution can explain war, dictatorships, cultures, and why some human behavior patterns are successful while some others are not.

We need to know why we do the things we do, and evolution can provide us with an understanding of human behavior.

We have yet another example of the misuse of education, and it has a direct and palpable impact on one of the most serious ongoing problems in the world today.

Ignorant people are rarely skeptics. We will see this pattern time and time again in Africa, South America, Asia, and in the Middle East. It is easy to gull ignorant people with grandiose claims that they have neither the intellect nor the sense to dispute. The main reason the Middle East is as explosive as it is, is due to the relative ignorance of the bulk of the population. Many of the Arab leadership claims would be open to dispute if the masses had the knowledge to question them. One need look no further than at the instruction of Palestinian children to see that miseducation is as bad or even worse than no education.

A Palestinian mother appears on TV, proud to have sent her eighteen-year-old son to kill both himself and "as many Jews as possible." What kind of woman would do this? What kind of cultural context would make this an admired act?

The answer is simple, straightforward, and quite human. It is a culture of hate. It is deep, visceral, malevolent, and entirely too human. Palestinian children are fed a diet of hate in school, at the mosque, and at home. The Holocaust is defamed, the calumny of the need for gentile blood for religious reasons is claimed, and the myth of Jewish control of the world to the detriment of Arabs is articulated. But it is easy to teach children to hate. They take to it naturally, genetically, for at one time it was essential to hate. If there was food for one and two to share, hate made survival possible. It was the survivor who left offspring and a predisposition to hate.

So ignorant, gullible, superstitious men and women go to their deaths serene in the knowledge that they will be looked up to at base camp. They will have status on Earth and seventy-two virgins in heaven. It is enough to make any young Palestinian with no visible future giddy with anticipation and eager to strap on the bomb.

Palestinian culture is a culture of hate. It is predicated on the Koran, which prophesies the worst of fates to unbelievers and authorizes the destruction of anyone who does not accept Islam. The Koran gives the Islamists an excellent and justifiable base from which to provide a religious hate to anyone who is not like "us" and a justification for violence.

But Palestinians are not alone. Hindus hate, Catholics hate, and Protestants hate. It is easy to dislike in the name of religion. I am right; you are wrong, and I not only have the right but also the responsibility to convert or kill. This is the philosophy of religion. It is also a true reflection of human nature. This is why we must insist on secular communities.

A secular country is a strong country. At first blush, this may appear questionable; after all, religion is supposed to be the ethical blueprint for living. Among Christians and Muslims, the Bible and the Koran supposedly elucidate a moral lifestyle although the specific response to a specific circumstance can vary depending on which chapter and verse is cited. Religion, we are told, is the guidebook for man. It separates us from the other life-forms on Earth and even provides a motivation to act as certain religious experts desire. And then there is always hell to which all sinners are banished.

Countries can be defined not by how they treat their majority but by how they treat their minorities. Most countries have an official religion. Others are characterized by a specific religious preference. England has, as its official religion, the Anglican Church. Poland is Catholic. Spain is Catholic; so is Mexico and most of Central and South America. Italy is very Catholic while Israel is Jewish, and Syria, Egypt, and Jordan are Muslim. The list goes on and on. Even Russia, that great atheist bear, is, at its base, Russian Orthodox. China does not have an official religion save communism, but both Buddhism and Taoism have deep roots in the Middle Kingdom. Even the United States has a Judeo-Christian background but with a twist.

The United States treats its minorities the same as its majorities. This mandate is found in the first part of the First Amendment to the Constitution that says, "Congress shall make no law respecting an establishment of religion or prohibiting the free exercise thereof." The state does not establish a religion, nor does it prefer one religion over another. The church and state are separate. As a result, religions, even minority religions, have the same rights and benefits as a majority religion. This is one of the aspects of the United States that makes it such a great nation. Because the United States does not discriminate against any religion, individuals from all religions have an equal chance to contribute to the state and receive benefits from the state.

There are some very practical problems associated with imposing a religion on a country. Islam presents us with an excellent example. Many Islamic countries governed by the precepts of the Koran prohibit the loaning of money with interest, proclaiming it is against the law of the Holy Book. This is simply not true. The Koran is specific. It states in Surah 2:275 that the charging of excessive interest is prohibited. The word that is used is closest to the word *usury* in the English language, which means the charging of excessive interest. In typical religious fervor, the leaders of Islam have decreed—because the Koran did not define usury—that all loaning of money at interest is prohibited. The economic result of this religious prohibition is truly monumental, especially in this day and age.

The ownership of homes and property is one of the defining characteristics of the United States. We have a huge middle class— rich beyond measure, having trillions and trillions of dollars, the single greatest accumulation of wealth anywhere in the world. Houses, town houses, condos, apartments, cabins, summer homes, winter homes, house trailers, and motor homes are all freely available to most of us, but only possible by the eternal mortgage. Money borrowed at interest in order to finance a home and over the years as more and more of us buy and occupy a home of our own, the value of our individual wealth goes up, the value of our country's wealth goes up, and prosperity blooms.

This is not so in true Islamic countries. If the law of the land is Sharia, then the only way to own property is to pay cash or inherit. Either way, both the wealth of the individual and the wealth of the country is seriously handicapped.

From an economic standpoint, Islam has a huge problem. But the culture of Islam is not just about money. It is also about face, and saving face can be a cruel taskmaster.

There are no more truly primitive cultures in the world. Clans like Black Rock or Lakeside no longer exist and have not for eons. Simple hunter-gatherers exist no more, and the reason is intelligence. For a long time, we had intelligence but used it only fitfully. We were the top predator in Africa, Europe, and Asia. We had fire, we had fine stone and bone and wooden tools, but most of all, we had intelligence, and intelligence began to color our culture.

At one time, when a female came into heat, she was serviced by all the males. The process was accomplished by a female resting on her hands and knees and the male approaching from behind. No connection was made between sex and procreation. There was no knowledge of procreative potential from a male viewpoint and little or nothing in the way of male parental involvement. The women had the kids; the women raised the kids, and as far as the male was concerned, that was just the way it was.

Then the male found out that he was as important as the female to the procreative process. If it wasn't for him, she could not have any kids. Males had, at this time, the intelligence to appreciate this knowledge. They responded by developing male-dominated cultures. This was not the only force causing our ancestors to begin to form larger, more complex, and more structured living conditions. But it was one of the original causes of more complex cultural systems.

As a result of this knowledge, monogamy was forced upon the female by the male, and it never really took. Cultures evolved that included, as a primary consideration, the establishment of a male-dominated society, one of whose objectives was to ensure the paternity of offspring. The woman always knew. She knew with whom she had had relationships. The male didn't know. He knew that he had a relationship but not if others also had had a relationship with the female.

Cultures are dominance systems. Where cultures—as part of the dominance pattern—consider precise male paternity a key component of that culture, we will see attempts by the males to control and protect paternity. The males will impose monogamy upon the female and will develop cultural reasons for and cultural ways of imposing that monogamy. Sometimes it is not pretty.

Cultural imperatives can be cruel, inhuman taskmasters. Cultural commandments can be a curse, and when male pride is at stake, women can be stoned.

Her name is Zafran Bibi, and she is in a Pakistani jail, sentenced to be stoned to death because she was raped. Zafran's story is typical of the way women are treated under Islam. Her husband was in jail for murder. She continued to live in his house, and his brother repeatedly raped her. She complained to her in-laws, but they did nothing. The rapes could have gone on indefinitely because under Islamic law, the only way a woman can prove she was raped was to have four Muslim men of upright character testify that they saw the rape. The woman's word is not admissible, nor is another woman's word or the word of anyone who is not a Muslim. It is impossible for a Pakistani Muslim woman to prove rape in Pakistan, which means there is a lot of rape in Pakistan.

The brother-in-law would take advantage of Zafran at his leisure and was getting away with it, too, except Zafran got pregnant. When the baby was born, it was proof positive that she had had sexual relations with someone not her husband on account of he was still in jail. Under Islamic law, that was all that was necessary to know. The circumstances were not important. The verdict was guilty. The punishment for her was to be tied to a stake while males, even the male that raped her, threw stones at her until she died.

This type of behavior toward females is common in Islamic countries. Men have rights; women have none. Men can come and go as they please, but women cannot go anywhere without a male escort. Men can wear what they want, but women must be veiled, cloaked, and wear a chador. And the question must be why. What possible good can it do for a culture to deny literacy to one half of its population? What benefit accrues to cultures who treat females not as second-class citizens but as third- or fourth-class citizens? Why do these people do this?

The answer is actually very mundane, even ordinary. It has nothing to do with females and everything to do with male perception. Before we delve into the original male chauvinist pig genre, let us take another trip to the ugly side of Islam.

Her name is Waris Dirie, and she is a supermodel. She is also Somali, and at the tender age of five, she was circumcised. This is a particularly painful and debilitating procedure performed on some two million Islamic girls each year. The operation is done without anesthetic. It is done by old women using rusty razor blades, old knives, or even sharpened stones. At the very minimum, the hood of the clitoris is removed. For other Somalian girls, infibulation is performed. Under this procedure, the clitoris, the labia minora, and most of the labia majora are removed. The wound is then sewn shut, leaving only a tiny hole through which the victim can urinate and through which menstrual blood can flow.

The result is painful urination and a difficult menstrual period. And of course, as a result of this brutal painful procedure, the female, for the rest of her life, will not enjoy sexual intercourse. On the contrary, both sex and birthing will be hard, bloody, and painful, far beyond that which a normal woman must endure.

Do we begin to see a pattern here? Engage in any kind of sex other than with your husband under any circumstances and you die. You are to be accompanied by adult males at all times to ensure that you do not engage in sex with anyone except your husband. Wear hoods, habits, and veils to prevent anyone from seeing you. Wear baggy clothing to hide your shape, and if that is not enough, just excise your external genitalia to make sex painful, unpleasant, and difficult. It certainly looks as though the male half of this culture is doing its damnedest to prevent the female half from engaging in any kind of sexual conduct with anyone except her husband.

And that is exactly the case. In Muslim cultures, to be a cuckold is to be at the bottom of the cultural structure. On the other hand, if a Muslim male has a reputation for getting other men to raise his children, he is accorded status, looked up to, admired, and disliked only if it is the disliker's wife that he has been dallying with. Muslim males absolutely hate the idea of raising someone else's children. It is anathema. It just will not be allowed. As a result, it is the females who face the brunt of the laws and rules and edicts governing sexual behavior. From the Muslim males' standpoint, this is an ideal situation.

His name is Mamhet Ali, and he is vaguely uneasy. This is not supposed to happen. Mamhet is sitting in his favorite teahouse, trying to enjoy his 10:00 AM cup of tea. It should be a pleasant time, a time of small talk among friends, a time to catch up on the local gossip and perhaps to thumb through the latest edition of the newspaper, even if it is already five days old. It's just not right.

The source of Mamhet's vague uneasiness is Mohammed Farah, an oily man, lean and smooth in his manner and movements—a man with a reputation. Women, it seems, are somehow attracted to this kind of lean and smooth. *Suave* might be a better word. Clean, yes, and immaculately clad with only a bit of ostentation in his turban and polished shoes. Mamhet's unease grows. The man is making directly for his table; he would not get to finish the paper.

The oily one made a motion with his hand, sweeping across the chair opposite Mamhet's, asking without words to join the other. A good Muslim man is courteous, and so Mamhet nods ascent, and the man glides smoothly into the chair.

"And how is your wife?" asks Mohammed. "And your family?" he quickly adds. To ask after another man's woman was very bad manners. Before Mamhet could manage more than a startled gasp, the other, realizing his error, quickly continued, "And your oldest son, seventeen now, isn't he? Yes, and I hear that he knows the Holy Book by heart, truly an amazing feat for one so young."

Two conflicting emotions grip Mamhet—rage over the other's impertinent question and pride in his son's accomplishments. Pride wins out.

"Fine, the boy is fine," he mumbled. "A scholar now, perhaps one day a cleric." The thought fills Mamhet with a sense of satisfaction. To have a cleric in the family brought status and esteem to the clan. It would be a first. His people had long been farmers and herders, occupations not known for scholarship. Mohammed nods sagely. "Truly Allah—blessed be he—has smiled upon you. What more could a man ask for? A scholar is a source of joy for any family. You are to be commended, Mamhet Ali. You have provided fine guidance for the young man."

The vague unease was turning into a growing suspicion in Mamhet's mind. Mohammed made a couple of attempts to change the subject, but he could see that his faux pas had poisoned the atmosphere to a point that it would be futile to continue. Mohammed stood up, made his goodbyes, and prepared to leave. A small smile played around the corners of his mouth, a smile that Mamhet could not fail to see.

And so Mamhet was left sitting, tea and paper forgotten, black suspicion clouding his mind. Thoughts, unbidden and dreaded, came tumbling into his brain. His scowl deepened.

It had not occurred to Mamhet before, but he had met Mohammed Farah through his father-in-law. The man had known his wife's family long before he had. Could it be? Mamhet didn't think so, but now the possibility crowded into his thoughts. True, the boy was born a short nine months after the wedding, but Mamhet had quite proudly put this down to the superior quality and vigor of his seed. Now he wasn't so sure. Damn, damn, damn. The boy looked like his mother, and everyone knew that shouldn't be. It was the man's seed that grew in the woman. She was only the vessel. But whose seed? Damn and double damn. That was the rub. Whose seed? And only the woman knew. That was the real problem. A man could think it was his seed, but only the woman knew for sure. It wasn't fair. It just wasn't fair a man never knew for sure, never. Only the woman knew.

Mamhet was a devout Muslim and not one to question the wisdom of the Creator, blessed be he. But in this one instance, surely a better arrangement could have been designed. Not that the man would carry the baby, but somehow in some way a man needed to know for sure. The Taliban had a good idea. Keep them barefoot and illiterate. Make them wear the burka. Forbid contact with males, any males except fathers and brothers. That kept them in their place and made a husband a little more sure that it was his seed. Mamhet had also heard of a practice that some Muslims used called female circumcision. He had been told that circumcised women got no pleasure from seed planting, even that it was painful. That would certainly keep them in their place.

Filled with doubt and utterly despondent, Mamhet went home, beat his wife, kicked his kid, and lived in a blue funk for the next three weeks. It isn't easy being a Muslim male.

This is why Muslim men treat Muslim women the way they do. It is cultural, it is dominance, and it is all status. It has nothing at all to do with taking care of females and looking after their best interests. It's a man thing.

Doubt is a terrible thing for a Muslim male. Doubt leads directly to hate, and it is hate caused by doubt that leads us to another Islamic problem. The reason many Muslims hate the West is because we cast doubt upon their most fundamental religious beliefs. Muslims are convinced, deeply convinced, that they are the one true religion. They and only they have a pipeline to God. They and they alone have the ear of the almighty.

And yet deep inside each Muslim breast, there is a tiny seed of doubt. If they are indeed the true people, the only people, why does God permit them to live as they do while treating the unbeliever to such a profusion of wealth, power, and dominance? Why has such a benevolent deity served them the plate of beans and bitter herbs while serving the infidels succulent roast lamb? And why, if God is on the side of Islam, have Muslims never won a war with Israel? It is too much for many Muslims. We make them doubt the only unique thing they have; even praying five times a day won't help. We should not be surprised that they dislike us so much. To show even the slightest doubt in God, to even suggest the possibility that the almighty is not is to risk hell. And besides, hating feels so good.

For a real understanding of human nature, we must look at the reality of religion. Religion is touted as one of the crowning achievements of man. It can be said—I think, without doubt—that humans are the only religious animal on Earth. Supposedly, religion is a sign of the moral superiority of *Homo sapiens*. After all, dogs don't go to church or to heaven.

But when we look closely at religion, troubling flaws mar its seemingly pristine edifice. The Catholic Church is the original Christian organization. For centuries, it held undisputed religious sovereignty over Europe. The original Christian philosophy as enunciated by Jesus was for a kind, loving, merciful, just, and forgiving religion. Jesus helped people, Jesus preached nonviolence, and Jesus preached justice. The Church turned the teachings of Jesus on their head and called the process an act of faith.

The Catholic Church came to power in the first half of the first millennium. By the first centuries of the second millennium, the Church was feeling its oats. In 1084, Pope Gregory VII was fighting Holy Roman Emperor Henry IV. The pope won but at a terrible cost—the sacking of Rome. Thousands of people died, and thousands more were sold into slavery, and both of these men were Catholic. In 1191, during the Crusades, English king Richard I, after capturing Acre, slaughtered three thousand helpless Muslim prisoners of war. King Richard I was a Catholic.

The Catholic Church, once in power and sure of itself, had no qualms about destroying any hint of a challenge to its dominance. In the very early 1200s, a number of individuals in Southern France set up an alternative Catholic Church. It was called the Albigensian Movement, and its creed differed substantially from the mother church. For one thing, they were opposed to church corruption, as common then as it is today. They also rejected the Old Testament, the sacraments, and most of all, they rejected the priesthood. The Church would not have its dominance questioned, and so the Albigensians died.

In the thousands they died horrible deaths at the hands of Catholic armies and the Catholic Church. One soldier asked how he was to tell heretics from faithful Catholics. He was told, "Kill them all. God will sort them out." It took twenty years, but in the end, the Albigensians were wiped out—man, woman, and child. This was done by a church whose Ten Commandments include "Thou shalt not kill."

It is 1478. The place is Spain, and Catholics will kill again on a massive scale. They torture huge numbers of people, murder in the name of God, burn thousands at the stake; and again, the Catholic clergy calls it an act of faith. This is the Spanish Inquisition. Its first priestly leader is Torquemeda. Its goal is the conversion or expulsion of everyone in Spain who is not Catholic.

The Inquisition went on for almost three hundred years, swallowing up hundreds of thousands of people and is as black a stain as can possibly be made by a religion on a totally innocent people. The Inquisition was not just death, but horrible, cruel, and pitiless death. Members of the Church, of the clergy—in the name of Jesus Christ, a gentle, forgiving, loving soul—slaughtered hundreds of thousands of people who did not see things exactly as the clergy did. It is dominance—simple and extreme. Believe as we do, acknowledge us as your masters, or die.

The atrocities of the Catholic Church did not stop with the Inquisition. In 1564 in France, a group of Protestants called the Waldenses were attacked by the French Catholic king Francis I. Thousands of Waldenses were slaughtered by a Catholic rabble. Christians fighting Christians. People who worship the same God, use the same book, and have the same savior butcher each other because they cannot agree to be different. Sounds like Northern Ireland.

The history of Europe has been a history of war, and almost every one of them was religious in nature. The Thirty Years' War and a host of others were all fought around religious issues. Millions of people wounded and killed in the name of God—my god, not your god, and if you doubt, you are dead even if one of the key tenets of my religion is not to kill. It is dominance, it is human nature, and it is reality. We need to know and to understand how this can happen. It will only stop when we truly grasp the true nature of man.

But it is not just the Catholic Church behavior in the past that must be examined and understood. We have talked about the problems in the countries of South America, Central America, and also Mexico. These are all poor countries with a small highly organized elite at the top and a huge peasant underclass. They were discovered and colonized at the same time as North America. They, however, developed in a totally different direction. Today, these countries wallow in corruption, greed, and fraud. In some countries, 20 percent of the population own 80 percent of the wealth while the great masses are regulated to a life of abject poverty.

The Catholic Church is the primary reason these countries are the way they are. In concert with the powerful, the Church developed a philosophy of subjugation and domination in order to maintain its control over the large mass of the underclass. The Church held all the keys; they could give or withhold education. Mostly they withheld. Ignorant, gullible, superstitious peasants are easier to manipulate. The Church could give or withhold salvation, and salvation was dependent upon the good graces of the Church. And the

Church said obey the master and us. The Church was the connection between the elite who had everything and the rabble who had nothing. The Church was supposed to be objective and an honest broker. But the Church was in bed with the elite, for that is where the power lies and that is what the Church feeds on—power.

The following story may or may not be exactly true, but what happened closely represents what happened thousands and thousands of times in Central and South America.

This is the story of a poor man named Pedro. Now Pedro has a surname, but it is immaterial and not important to the story. Pedro lived in South America during the 1700s in Argentina or Chile or Brazil or some such country. As I said, Pedro is poor. He lives by sufferance on the estate of the patron, a large florid man given to afternoon siestas and huge lavish banquets. Banquets to which Pedro is not invited nor even thought of. Indeed, sometimes it gets so bad that Pedro's belt buckle meets his backbone, and his belly cries for beans. Pedro, when he is hungry, thinks of these banquets where he barbecues the patron's beef but is prohibited from partaking of his own cooking. It is totally unfair, but what is a poor, unlettered peon to do?

He must work for the patron to maintain his position. He lives in a tiny dark hut along with his wife and two children. In what little spare time he has, he tends a small garden and two goats. Pedro's life, while not the best, is better than a lot of his contemporaries. The patron has a terrible temper, and many a friend has been ordered off the land at the slightest whim. It is a death sentence for most, and they either starve or are murdered by bandits.

Pedro does have something he is proud of though—his children, a boy, twelve, and a beautiful daughter, just fourteen. Both bring tears of pride to Pedro's eyes for both (out of an original total of seven) have gone from being good children to being good young adults. Both work hard—the son in the garden and with the goats, the daughter her mother's perfect helper. She spends her time grinding corn, washing dirty clothes, sewing, and the hundred and one absolutely essential chores necessary to keep the household together. By any standard, Pedro is poor, but with his loving wife and well-behaved children, Pedro was actually able to count his blessings.

One day, while Pedro was out working at fence mending, the patron's son and a few of his friends rode up to the hut. The young aristocrat had seen Pedro's daughter at market the week before and meant to have her. The mother interfered and was badly beaten. Pedro's son tried to stop the kidnapping and was stabbed through the heart. The girl was taken away by the gang.

Pedro came back that evening to find his daughter gone, his son dead, and his wife badly beaten. In one fateful afternoon, Pedro's life and purpose fell apart. His daughter, too ashamed to come home, became a prostitute in the town and was killed a year later for the few pesos that she earned at her trade. Pedro's wife never recovered from the beating she received and spent most of her days in the corner of the hut, rocking back and forth and crying out for her lost daughter.

Pedro was not entirely alone in his misfortune, however. The local priest, upon hearing of the crime, hurried to the hut to console the distraught and depressed man.

Pedro could not understand. He was a good Catholic; his wife and children were good Catholics. How could this happen? Why did this happen? Pedro was grief-stricken (he knew better than to be angry; any intimation that it was the fault of the system and the patron would lead to eviction) and desperately needed the cleric to explain how this could happen to a poor but Christian man.

"It is for the best," intoned the priest. "God works in mysterious ways, the goals of which we cannot understand but are for the best, else they would not happen, for God lets nothing happen without his approval."

"Yes, but…," stammered Pedro, "why?"

"You must give yourself into God's hands and trust for the best," pointed out the priest. "After all, you, Pedro, were made in the image of God. Think of what that means. In this time of grief, put your trust in the Church and in God. Everything is for the best even when it does not appear to be so."

Pedro found it hard to follow the priest's counsel. The idea that the patron's son could ruin a man's family without punishment was hard to swallow. But Pedro was poor, ignorant, gullible, and superstitious while the patron was rich and the priest was educated. What's a poor man to do?

Things have changed in the Church, and yet the motivation is still there. No more Pedros or at least not as many, but the behavior of the priest still reflects the true goals of the Church. The leaders of the Catholic Church follow today in the comfortable, familiar shoes of the leadership of the past. Today the Church owns billions and billions of dollars of stocks and bonds and property. The Church also has vast treasures of gold and silver and jewels and paintings and books and artworks and all kinds of very, very valuable treasures.

Now Christ was poor, and he preached poor. Wealth would not help in the kingdom to come, only good deeds. If the church truly believed in God and their own creed, they would sell everything they own and give the money to the poor. They would do this secure in the knowledge that God would provide.

Don't think it's going to happen.

The Church has adamantly opposed any kind of birth control and equally opposed abortion. Yet over fifteen thousand children die in the world every day from causes related to nutrition, causes that could be alleviated for less than a dollar a day. But the Church likes the power that comes with wealth and the benefits related to having wealth. No moldy crusts of bread on a tin plate for the prelates but rich, hot, and tasty victuals served on fine china or even silver. Oh, to be rich, to drink fine wines, to be served by valets and butlers—this is heaven divine. The choice is simple for the Church: "Let 'em starve."

That is not to say there are no good Catholics. One individual comes to mind, one who does indeed act the way Christ taught, living a poor humble life, helping the needy and the downtrodden.

This individual has a history of giving unstintingly and consistently, living simply, modestly, and charitably. In short, the life Christ expected every Christian to live—following his teachings and doing what he would do.

The interesting thing about this particular individual who did just what Christ would do is that her behavior was so unusual, so unique, so extraordinary that she qualified for a Nobel Prize. And yet she was only doing what any good Christian should do, was supposed to do, was taught to do. In Mother Teresa, the difference between Christians and the Church become apparent.

Mother Teresa followed the teachings of Christ to the letter. The Church follows the genetic predispositions of dominance. One man, one individual at the top—supreme, infallible, and without a rival. Feeding and maintaining that delusion underlies the entire structure of the hierarchy of the Catholic Church. Rich robes when plain garments would do and the poor would benefit. Gold goblets when plain pewter would do. Silk sheets when cotton would do. And all that savings could go to help the poor. But it is not to be. Rich, powerful men (and they are all men) must have the accoutrements of power. The masses must bow in awe before the richness, the grandeur, the absolutely overwhelming affluence of the dominant. Let the poor go to hell.

Money is power; wealth is dominance. And the Church covets dominance. Control and authority at any cost is the order of the day. We need look no further than the recent episodes with the pedophiles in the priesthood. Everything was done to cover up the scandal. Known child abusers were routinely shipped from parish to parish when rumors of their behaviors started to surface. No thought or concern was ever given to the thousands upon thousands of children permanently scarred by being violated by a person in a position of authority. This is the real gospel of the Catholic Church, protect the priest at all costs. Nothing must be allowed to challenge the absolute authority, the papal infallibility of the Church. It is a power thing.

It is not just the Catholic Church that presents us with a reality totally divorced from religious conviction. Islam is, today, a religion of death. In the Middle East, in India, in the Philippines, in just about any point on the globe that has any kind of Muslim population, we have death and destruction. There are exceptions, but they are few and far between. Muslims seem to get a great deal of pleasure out of death. One of the few populations celebrating 9/11 were the Palestinians. Newsreels (but not very many because the press did not want to give the wrong [right] impression) showed the people of Gaza and the West Bank celebrating the collapse of the World Trade Center towers. These Muslims were ecstatic over the deaths of three thousand innocent individuals. Individuals who couldn't fight back, who couldn't defend themselves. Individuals who did not even have the tools or training to fight back. This seems to be a defining characteristic of Islamic terrorists. They always pick on civilians, on the defenseless, on those who cannot fight back. They recently had the intestinal fortitude to shoot a man who was confined to a wheelchair and then push him off a boat. This takes real courage.

We see this kind of behavior in most religions, the basest of acts for what are claimed to be the highest of motives. It is the motives of power and dominance under the robes of piety. It is the hatred of men in pursuit of the domination of mankind. Religion is not a good thing. On balance, it has cost millions and millions of lives and continues to do so today. Where countries sanction a specific religion, death and destruction mount. It continues today in India, Somalia, and a myriad of other countries. Religion is not a good thing. Separation of church and state is absolutely essential.

Chapter 16

The fifth and last requirement for an ideal country is that it be fair and just. Everyone must be treated the same; everyone must be subject to the same laws. Laws must not be biased and must be uniformly enforced.

Everyone must have the same chance at success within the bounds of reality. It is just as bad to insist that a below-average student be admitted to a prestigious institution of higher education as it is to deny entrance to an above-average student based on race, creed, or sex, or any other arbitrary category. Student acceptance must be based on ability and desire. There can be no favorites, no exceptions, and no quotas.

What is true for students is also true for everyone else. Laws must be uniformly applied, rules equally enforced, and the rule must be realistic. The concept is fairly simple, and it is enshrined in the Declaration of Independence of the United States. Everyone has the right to "life, liberty, and the pursuit of happiness." It is as simple as that.

Notice it does not say life, liberty, and happiness—only the pursuit of happiness. How, when, and if one reaches happiness is left up to the individual, but happiness is not guaranteed.

Here in the United States, we come as close as any country has in providing the opportunity for everyone and anyone to achieve happiness. We do not have castes, we do not have classes, and we do not have royalty. As we shall see, it is because these conditions do not exist that America is truly the land of opportunity.

India is democratic. India is capitalist. It is not as secular as it could be, and as a result, Hindus and Muslims murder each other on a regular basis. Education is also a problem in India. The country does not educate the vast bulk of its children, and that is a real weakness.

But the most egregious failure in India lies in its caste system. Here we have the very worst of human nature, writ large and deeply embedded in the Indian culture. There are five basic castes in India, and one's social position is decided at birth. First come the Brahmans, the elite with the bulk of the power and money in India. They like the way things are in the country and are not above committing illegal acts to keep things that way. Next come the Kshatriyas, the warriors and rulers. Third are the Vaishyas, a caste of farmers, merchants, and artists. Fourth are the Sudras, the laborers. Last and least are the untouchables, those at the very bottom of the social, cultural, and economic ladder. They are the lowest of the low, the least of the least, and the reason India festers with injustice and bigotry.

The fetters of the caste system in India have gotten so bad that some of the laborers and untouchables have changed religion. It is the Hindu religion that mandates castes and the social status of each individual. To get out from under the pall of the caste, many Indians are converting to Islam or Christianity. It is getting so bad in some areas that the local government is passing laws prohibiting the switching of religion without the approval of the civil authorities. The Brahman must have their untouchables to validate their superiority. It makes them feel so good to know that they are so much different from the others.

India will never be a great country. Any social group that marginalizes and discriminates against half of its population will never succeed. So even though India is nominally a democracy and capitalist, it will never be able to provide the greatest benefit to its entire population as long as the Hindu caste system is enforced.

Mexico, our great neighbor to the south, today demonstrates the value of a fair and just milieu with respect to economic models. They also demonstrate just what happens when human nature is not constrained by cultural mores. Mexico, as I have said, was discovered at the same time as the rest of North and South America. It was settled by Europeans. It had vast natural resources, a varied climate, and tremendous potential just as the rest of North America had. But today Mexico wallows in crime, corruption, and greed. It is controlled by a tiny elite—highly motivated, highly organized, and interested in only one thing, dominance. It is run, literally, by men behaving badly but normally.

Our story must begin with the Mexican government, for Mexico's troubles originate with those in charge and their motivation. For hundreds of years, Mexico has been a country of dictatorships. Under the Spanish thumb, Mexicans were at the mercy of Castilian despots with no rights nor any rule of law. Needless to say, Mexico was not fair and just.

Even after independence from Spain, Mexico fell into the same rut as all the other Central and South American countries. Dictators and dominance, individuals running the countries to suit themselves, to make themselves and their cronies rich, to take advantage of political position to advance personal agendas, and to exact revenge upon real and perceived enemies. Until the year 2000, this same condition existed in Mexico—dictatorships, dominance, greed, and corruption.

The elite tried to put a democratic face on their dictatorship, trotting out a different Institutional Revolutionary Party (PRI) candidate every few years. But the result was always the same—loot the economy (one president's brother made well over a $100 million while his sibling was in office by peddling access to state power), repress any and all resistance (a student protest was viciously put down in 1968 with the state killing some three hundred to four hundred protesters), and, in general, run the country to benefit a small group of individuals at the expense of the great majority of

Mexican citizens. Graft, greed, and corruption are the order of the day, and this atmosphere is the prime reason Mexico today has the problems that it has.

It starts at the top. The PRI was recently fined $100 million by the Federal Electoral Institute of Mexico for using government funds to illegally finance its 2000 presidential campaign. The corruption would have been kept secret except that for the first time in seventy years, the PRI lost the presidential election to the PAN (National Action Party). The behavior is typical of the PRI and has been going on for its entire history. They'll do anything and everything to get into a position of power, of dominance. Anything and everything to stay in power. It is ordinary, expected, and endured by the populace because they do not know any better. The rot at the top also shows up at the bottom.

A young Mexican engineer needed to get a new driver's license awhile back, and how he accomplished the job is revealing. He went to the driver's license bureau, slipped the clerk $10, skipped the driver's test, and walked out with a new Mexican driver's license. When that same engineer wanted to get married, he used another bribe to get the wedding date he wanted. The citizens of Mexico pay bribes to get a car registered, garbage collected, children registered at school, water service turned on, and a vast range of other public services. Bribery is endemic in Mexico, and it goes from bottom to top. That same engineer that bribed his way into a driver's license and a wedding date found out that the construction company he worked for had to pay a $100,000 bribe in order to get a building permit.

Bribery is ingrained in Mexican culture. It occurs because the Mexican populace lets it occur, and it does tremendous damage to any relationship a citizen has with the bureaucracy. A bribe means that all are not treated fairly in their interactions with the police, the civil authorities, or even the judiciary. If Mexico is to truly become a great nation, it has to treat everybody the same. It has to treat everyone honestly and fairly, not with corruption and greed. Mexican citizens' willingness to put up with this kind of behavior—indeed, their actual preference for the bribe—will keep Mexico from ever becoming a wealthy country.

A country must not only be fair, but it must also be just. Here, too, Mexico fails miserably. The Mexican police are among the most corrupt in the world, and yet a country's first line of defense against chaos is an honest police force.

Just recently the Mexican army raided federal police anti-narcotics offices in eleven states in Mexico. The police who made up the unit were all placed under military control until the government could figure out who was corrupt and who was not. The mess had gotten so bad that the antidrug unit could not be saved. It was not a question of a few bad apples but a whole barrel full. The very government unit tasked with rooting out drug dealing was in bed with the dealers.

What we must realize is that this is not unusual behavior. A police department that displays moral and honest behavior is unusual. It is not a case of a few rogue cops but rather of men behaving normally. It is our perception that needs to be changed. It is the only way we will truly come to understand man.

This is not to say that honest police departments do not exist; they do. England has one of the best. Bobbies have the reputation for being fair and for treating all equally. Once in a while, a royal will get off while an ordinary might get a ticket, but that is the extent of the favoritism. Great Britain's police are among the finest in the world. It is because they are expected to be. English culture demands a fair and just police department. Corrupt cops are looked down upon, treated with contempt, and when caught, punished. It is English culture that ensures that England's police behave honestly. Because English culture demands it, only individuals who are honest and upstanding are allowed to become police officers. If once in a while a bad apple slips through, when caught, he (or she) is quickly removed.

In Mexico, the police department is a political machine. Membership in the police is as much based on political pull as it is on ability and training. Many cops are simply the goon squads for the PRI. They are given police department jobs for the express purpose of shaking down the populace and keeping political enemies in check. Mexican culture, too, contributes to the problem. Mexicans who disobey the law and have money find that a *la mordida* is a quick and easy solution to running a stop sign or speeding.

The corruption in Mexico is endemic. It permeates all levels of society and politics. Every office is for sale, and every government-run business is a cash cow for its political appointees. This situation can be changed. Corruption can be stopped or at least greatly reduced. Bribery can be brought to a standstill, and even the looting of the country by the politicians and their cronies eliminated. But it will require a wholesale change in Mexican culture. Mexican political thieves get away with murder because they can. They loot and misuse state funds because the Mexican people let them. If the people of Mexico want honest, law-abiding, and fair leadership, they must demand it. They must fight for it, and they must socially punish anyone and everyone who fails to behave honestly and justly.

There are five parameters for greatness in any country. The country must be democratic, capitalist, educated, secular, fair, and just. These fundamental requirements are necessary because of the nature of human nature. We must have democracy because it is the only way a country can control its alphas. Dominant individuals would prefer to reign unchallenged. Dictatorships of today such as Cuba and North Korea and of the past such as Iraq demonstrate this decisively. Alphas can be made to conform to society's demands. They will follow the dictates of the people but only if they are forced to. It is a component of human nature for some to lead; it is a genetic predisposition, it cannot be eliminated, and it can only be controlled.

If society is to provide the greatest good for the greatest number, that society must be capitalist. As we have seen, wealth is a matter of time. All raw materials are free, and it is only the labor that is applied to those raw materials that costs. Cost is determined by the length of time that it takes to convert raw materials into finished products. The less time it takes to do this, the more wealth there is. It is as simple as that.

There was a time thirty years ago when if you wanted to see a specific invoice you had to go to the city in which the invoice was filed, visit the building in which the invoice was filed, enter the room in which the invoice was filed, and talk to a specific individual. That individual would get up from a desk, walk over to a specific filing cabinet, and extract that invoice. That is, assuming that someone else had not already extracted that invoice or that the invoice in question had not been misfiled in the first place. It was a long, slow, labor-intensive process.

Then came computers. Today, you call a number, you give the file clerk a number, and the invoice appears on a computer screen. The saving in time and effort is enormous. Two points: First, the record economic run during the Reagan and Clinton administrations had little to do with who was president and much to do with businesses taking full advantage of the computer revolution. Computers have saved so many man-hours of labor in the last twenty years that raises could be given, production could be increased, accuracy updated, and control of inventory improved to a point that businesses could make a profit. At the same time, they could keep prices of goods fairly constant. Second, Bill Gates is castigated for his wealth. But Bill Gates's wealth is a mere pittance compared to the trillions and trillions of dollars the public has saved as a result of his computer programs. We should not begrudge him his billions, for he has saved us trillions. Likewise, we should not envy the Walton billions. They, too, have saved—by their marketing skills and ability—the American, indeed international, customers trillions of dollars. Sam Walton and Bill Gates are alphas, dominants. They thrive and are successful in the United States because this country is a capitalist country. Capitalism counts.

Education is the key. We cast our lot with intelligence more than a million years ago. It was not an easy choice. Teeth—sharp and long—are better. Claws—sharp and long—better still. But we had none such and had to make due with fire and intelligence long ago. Fire winnowed the intelligence wheat from the instinctive chaff fairly quickly. Those who are not our ancestors lost fire and died. Those who are our ancestors mastered fire and fathered us. It was knowledge that did it, learning that counted, and we survived and became top predators and then moved beyond predation.

We did it by learning, by exercising our brain, and in response, our brain grew. We learned more, and our brain got bigger, and we learned more. In a typical evolutionary way, our brain got to its present size and hit a roadblock. We achieved a balance between birth and brain. A larger brain meant a more difficult birth, and the selective process caused today's brain size. Our Neanderthal neighbors tried the "bigger is better" mode and failed. Neanderthal brain cases in adults are slightly larger than that of *Homo sapiens*. But Neanderthals were cursed with a bun, an extension of the skull at the back of the head. The Neanderthal baby's head, bun included, had to exit through the birth canal. While labor for today's mothers may at times be bad, among the Neanderthals it must have been much worse. A difference of only 2 or 3 percent in birth rates would, over a period of time, regulate the Neanderthal to the long list of extinctions.

Just as intelligence has kept us alive throughout our history, it is our only hope for our future. Everyone must be educated to the fullest extent possible. Everyone. Our future is one of overcoming the many crises we will face, and this can only be accomplished through intelligence and education. We face despots, and our salvation lies in knowing why they behave as they do. We face a rapidly increasing population with no end in sight. In the coming centuries, people will be fed by programs we can only dream of but will be possible with learning and understanding. We are faced with new diseases every day, and as population densities increase, we become ever more susceptible to epidemics caused and spread by these same high population densities. More people means more places for the disease to start, and more people mean an easier path for the disease to spread. Our future depends on learning, and everyone has to be educated to the greatest extent possible. Our survival depends on it.

All the world's countries must be secular if the greatest good is to be done for the greatest number. A favorite religion results in favoritism. Everybody must be treated equally from the greatest majority to the smallest minority. Where religious favoritism is manifest, where religious intolerance translates into discrimination, nations suffer. It is imperative that religion and government be kept separate if true equality is to prevail.

And the last but not least, we need to ensure fair and equal treatment for all. All the decay in struggling countries today can be traced back to unequal treatment of its citizens. Those in power and those who aspire to power can do no wrong. They lust after dominance and, behaving normally, will do whatever it takes to achieve their goals. If, in the process, individuals are harmed or killed or dispossessed or even unreasonably inconvenienced, so be it. This is ordinary human behavior. This is how alphas achieve dominance as long the populace lets them. Fair and just ensures that everyone plays by the rules and that the rules are the same for everyone.

It will not be easy. The United States is a fluke. We exist as we do today only because of the intelligence, common sense, and understanding of the Founding Fathers. Washington, Adams, Franklin, and a small group of wealthy landowners and business-men tried to set up a government that would allow them a good shot at getting rich and staying rich. In the process, they made it possible for anyone to follow in their footsteps. They could hardly do otherwise. Without a class system, without privilege based on birth, they had no choice. Everyone (white male property owners only) had an equal chance to succeed or fail because there were no other power groups in the new republic. There was no royalty, no patrons, no elite to which an inordinate amount of power could be granted. They were "just folks," and it is these people who started the ball rolling.

Today, the United States is the freest, wealthiest, most open country in the world. We are what we are because we are democratic, capitalist, educated, secular, fair, and just. It is a recipe for success.

Alphas are the key. They are always the key. For good or evil, they have an extraordinary impact on those they dominate. These are the same alphas that contested and schemed and blustered their way to the top of the clan, the group, the tribe, down through the eons of our survival on the plains of Africa. Present-day alphas have the same genetic predisposition our ancestral alphas had. They have the same desires, the same cravings, and the same willingness to satisfy those predispositions any way they can.

But this same desire can be used to control present-day alphas. Dominants will be with us always. In every generation, there will be born individuals with a genetic package that predisposes them to dominance. This cannot be prevented, but it can be controlled. Alphas will do whatever they have to, whatever is required in order to assert their predisposition. They would prefer no restraints upon their appetites, and when no curb is present, we have the likes of Stalin, Hitler, Mao, Pol Pot, Castro, Hussein, and a myriad of other two-bit tin-pot dictators both past and present.

But alphas can be made to conform within the confines of culture. The need is to lead; the predisposition is to rule. But leaders cannot lead followers who will not follow. It is as simple as that. Tomorrow, the Cuban people could free themselves from communism's iron grip simply by refusing to follow the current dictator. True, some would be killed, but the dictator and his minions would not be able to kill everybody, and as soon as it became apparent that the Cuban populace would no longer follow socialism, the leaders would be gone.

This is not just speculation. Russia today has many problems, but communism is not one of them. The Russian people put enough pressure on the communist dictatorship to force it into oblivion. East Germany was a dictatorship within a dictatorship. It, too, crumbled when enough East Germans decided to put an end to the Marxist farce. And that is the way it works. When the leaders find that followers will no longer follow within the parameters established by the leaders, then they must change the parameters or lose their leadership. This is how followers control leaders.

We are not naturally law-abiding. We do not, by predisposition, tend to obey rules and regulations. One way whereby rule of law can be implemented and maintained is where individuals risk loss of property when rule of law breaks down. This is one of the most important but least recognized attributes of private property.

People tend to adopt and live by rule of law when not doing so is going to cost them. When a person has a house or property or any other kind of asset, it is unwritten but understood that the assets only have value as long as rule of law is observed. It is no coincidence that riots take place in areas of high population concentrations in apartments. People who live under these conditions have little and so have little to lose (and much to gain by looting) in a riot.

It is hard not to be pessimistic about the future of man. We are the only species on Earth to have consciously created the means of our own destruction. We are the only species on Earth to overpopulate our planet to the tune of almost seven billion individuals. We cannot control our appetites, and so drug usage, both legal and illegal, run rampant. And there is always someone somewhere willing and eager to supply such addictions. We fail miserably to be responsible, and so a disease such as AIDS becomes an epidemic. Simple, easy, and safe procedures exist to combat the spread of AIDS. Do we listen? Do we take precautions? No. And so what should be a minor curiosity in the medical establishment is instead a major factor in life expectancy around the world.

The history of the world has been a history of war and dominance. In Europe, in Africa, in North and South America, in Asia, always it has been war and contesting and killing on a massive scale. Death and destruction writ large mostly for the sake of one man—or, rarely, woman. For one individual's desire to be emperor or king or reigning despot or even God. Our history is a litany of delusions on a grand scale, butchery of a monumental magnitude, awesome destruction—and for what? All this so that one individual could think a little better of himself. All this so an ego can be massaged. The sole reason for the deaths of millions upon millions of people is just so one person can feel a little more secure, a bit more important, of a slightly greater worth. This is human nature.

What is really appalling is the ease with which these individuals are able to accomplish their goals. In addition to our predilection for dominance, we seem also to have a positively amazing propensity for organizing such dominance hierarchies. No respectable dictator has any trouble at all rounding up enough sycophants to achieve the goal envisioned by the head man. Nobody says no.

We have discussed Stalin and his murderous reign—over sixty million dead—and the fact that Stalin did not, could not, have acted alone. Dictators must have their organizations, and it seems they never have a problem setting up a killing machine. To kill millions, you need tens of thousands; to kill tens of millions, you need hundreds of thousands. And no dictator has ever had a problem recruiting killers. Not Stalin, not Mao, not Pol Pot, not Castro—all dictators, all the heads of vast organizations dedicated to killing. It is another sad commentary on human nature when one sees how easy it is to set up a killing machine.

The solution is not all that difficult. It starts with the understanding of the real nature of the human animal. It requires a dispassionate, unbiased analysis of our behavior patterns. Our problems begin with dominance. Dominance has shaped and influenced our behavior for eons, as it still does today.

But it is with delusion that we must first deal. Delusion blinds us. We cannot see ourselves objectively. Our delusions are many and varied. In religion, we are just absolutely convinced that we are right, that others are wrong, and that we have not only a right but also a duty to bring others to our religion. There are three great religions in the world today—Hinduism, Islam, and Christianity. These three religions are mutually exclusive. Hindus believe in many gods; Islam and Christianity, only one. Christians view Jesus as the son of God; Islam says he was only a prophet.

So at best, only one of these religions is true while the other two are false. It is also possible that all three are wrong. Three-to-one odds are really not very good when it comes to killing and maiming to get to heaven. When we factor in all the world's religions—those that exist today or have existed in the past—the odds of believing in the right one climbs to tens of thousands to one. Terrible odds if one is a homicide bomber, and yet we cling to our cherished delusions. We are right; you are wrong. *Bang bang*, you're dead.

We see delusion in our politics, too, and fail utterly to recognize it. Again, we have the perfect example. Neville Chamberlain thought he had a pretty good thing when he came back from Munich in 1939 with a piece of paper. A piece of paper Chamberlain said would bring "peace in our time." In September of 1939, Hitler invaded Poland, setting off World War II with its years of terror and millions of deaths.

Chamberlain thought Hitler was—at bottom—a rational, reasonable man. Chamberlain knew that if he held his mouth just right and said the proper words, Hitler would see that—which any intelligent reasonable man would see—if both parties just try hard enough, things can be worked out without violence. Chamberlain did not understand, even though he was one himself, that we are always led by dominance-oriented individuals and that some of these individuals have huge appetites. Chamberlain was deluded. As a result of his delusion, tens of millions of people died.

It begins with education, with understanding, and an ability to see humans without the rose-colored glasses of humanity. It ends with understanding. Only when we truly understand human nature will we be able to deal with the human animal. It very well may happen in this way, but I am not optimistic.

What is it to be? Will we tame our demons and cage our delusions? Can we combine education with a healthy dose of skepticism to achieve a modicum of reality? I just don't know. Again, to the chagrin of the politically correct elitists, I will make a few inappropriate statements. The United States is the single greatest country in the world by a long shot and by any measurement. The United States is the most powerful country in the world. It is the wealthiest, the freest, and the most beneficial country in the world. If every nation in the world were like the United States, we would have no war, no poverty, no ignorance, and no (or very little) government corruption. The real problems are elucidated by the following: democracy beats dictatorships, capitalism defeats socialism, education overcomes ignorance, and just behavior conquers corruption.

We can also look at the United States as a model for creating a better world. As I have said, this is not a popular course for some who feel that the United States does not reflect the true nature of man. But in an imperfect world, we will just have to deal with the imperfect.

The question is can we?

We are not the perfect species; we are not the logical culmination of the evolutionary process. Bipedalism is a response, not an end. Intelligence is a survival mechanism and not a very good one at that, else we would see more intelligent animals. Consciousness is an accident and not the hallmark of a superior life-form. We are not special, just different just as each species is different.

We do, however, have the means within us to become a better species. We can stop war. We can ensure that everyone reaches his or her full potential. We can all be free. We have the means, but do we have the will?

www.ingramcontent.com/pod-product-compliance
Lightning Source LLC
Chambersburg PA
CBHW021420150726
47989CB00001B/45

9798887315898